S. William Beck

Gloves, their Annals and Associations

A Chapter of Trade and Social History

S. William Beck

Gloves, their Annals and Associations
A Chapter of Trade and Social History

ISBN/EAN: 9783337204037

Printed in Europe, USA, Canada, Australia, Japan

Cover: Foto ©berggeist007 / pixelio.de

More available books at **www.hansebooks.com**

GLOVES,

THEIR ANNALS AND ASSOCIATIONS:

A Chapter of Trade and Social History.

BY

S. WILLIAM BECK, F.R.H.S.,

Author of "The Drapers' Dictionary: A Manual of Textile Fabrics," &c.

In whose leaves
The very marrow of Tradition's shown,
And all that History, much that Fiction, weaves.

LAMB.

LONDON:

HAMILTON, ADAMS & CO., 32, PATERNOSTER ROW.

1883.

LONDON :

PRINTED BY W. H. AND L. COLLINGRIDGE, CITY PRESS,

148 AND 149, ALDERSGATE STREET, E.C.

CONTENTS.

Historical.

CHAPTER		PAGE
I.	ETYMOLOGY OF GLOVES	1
II.	ANTIQUITY OF GLOVES	6
III.	GLOVES IN THE CHURCH.	15
IV.	GLOVES ON THE THRONE	28
V.	GLOVES ON THE BENCH	51
VI.	HAWKING GLOVES	58
VII.	GAUNTLETS	68
VIII.	PERFUMED GLOVES	82
IX.	CHICKEN GLOVES	95
X.	GLOVES IN COMMON WEAR	103
XI.	COMPANIES OF GLOVERS	135
XII.	THE GLOVE TRADE	162

Symbolical.

I.	GLOVES AS PLEDGES.	189
II.	GLOVES AS GAGES.	203
III.	GLOVES AS GIFTS	227
IV.	GLOVES AS FAVOURS	252

INDEX.

PAGE

Anglo-Saxon customs'duties include gloves 13
————————— MS. shews gloves 13
Anne, S., patron saint of glovers 136
Antipathy, instinctive, of dogs to glovers 183
Antiquity of gloves 6
Appeal to wager of battle . . . . . : 210
————————— abolished : . 212
Arms of the London glovers 156
————————— Perth Incorporated glovers 139, 198
Articles of Association of London glovers 144
Asbestos, gloves made from 185

Bartholomew, Saint, patron of Perth
    glovers 137
Berlin gloves 176
Betrothal, legal, gloves binding token of . 236
Bishops' gloves, ancient 19, 25
————————— modern 17
Bishop, Child 22
Biting the glove, sign of defiance . . . 224
Braels 148
Bribe, gloves expressive of 54
———— gloves offered as 54, 249
Bucks' leather gloves 170
Bugle-trimmed gloves 129
Byssus, gloves made from 185

Cave men, gloves of 12
Challenge, gloves cast or sent in . . 78, 223
Champions 209
Charles I. gives gloves at his execution . 43
Charter of incorporation of London glovers 159
Chattertonian controversy 112
Cheverell gloves 115
Chicken gloves 95
Church, gloves in the 15
———— laity required to doff gloves on
    entering 16
———— colour of gloves changed with
    seasons 17

PAGE

Church, gloves hung up in challenge in . 225
Clergy forbidden to wear gaudy gloves . 20
Close gauntlet 78
Coronation, gloves employed in 29
———— place of gloves in ceremonies . 38
———— challenges at 212
Cypher gloves 131

Dannocks, hedging gloves 180
Degradation, the gloves taken off . . . 201
Derbyshire churches, funeral garlands in . 244
Dogskin gloves 181, 183
Domestic manufacture of gloves 180
Double gloves 114
Drawglove, a game 234
Durham Churches, funeral garlands in . 243
Durham, Bishop of, escape from Tower . 21
Duty, Customs, on gloves . . 164, 169, 171
———— on gloves repealed 173

Easter gloves 230
Edward I. opening of the tomb of . . . 36
Edward IV., gloves bought for 40
———— burial of 35
Elizabeth, Queen to Henry VII., burial of 34
Elizabeth, Queen, her gloves 42
———— gloves given to . . . 49
Enamelled gloves 28, 184
Etymology of glove 3
———— gauntlet 69
Execution of Charles I., gloves given to
    his attendant 43
———— Mary Queen of Scots . . 43

Faith typified in gloves 3
Fairs, former importance of 193
—— gloves sign of security at . . 191, 194
"Fair Maid of Perth," her house . . . 137
"False" gloves burnt 148
Fashions begun in defects 87
Favours, gloves as 252

PAGE

Fees on taking up freedom of glovers'
company 160
Flag of incorporated glovers 139
Foreign gloves, import of prohibited . . 162
Foreigners, trade with . . . 144, 152, 154
France, companies of glovers in 136
Frangipanni gloves 92
Frauds and faults in gloves provided
against 145, 148, 152, 159
Free trade in gloves 174
Fringed gloves 132
Funereal garlands 240
Funerals, gift of gloves at 245

Gadlyngs introduced 72
Gages, gloves as 203
Garland, funereal 240
Gauntlets 68
Germany, early manufacture of gloves in 103
Gifts of gloves, 41, 67, 83, 84, 85, 114, 155,
228, 247
—— of lovers 3
Gloves to be dispensed with 133
—— worn at night 95
—— "of the good governance" . . . 110
—— carried in the hand 112
—— exported, *temp.* Rich. II. . . . 155
Glove money 230
Glover, surname 141
Graves, gloves placed on 243
Greeting, gloves doffed in 56, 134
Greeks, gloves of 10

Hand and glove 197
Hand to be ungloved in courtesy . 56, 134
Hands, models of, made to fit gloves on . 132
Harvest gloves 231, 248
Hawking gloves 59
Hedging gloves 117, 180, 248
Henry, Prince, son of James I., his gloves 41
Henry II., burial of 34
—— his effigy bears gloves . . . 35
—— VI., gloves of 39
—— VII., gloves bought for 40
—— VIII., gloves bought for 41
—— his hawking gloves . . . 63
Horse-hides, gloves made from 134

Illicit trade in gloves 162, 166
Import of gloves prohibited 156
—— permitted 174
Inch of candle, auction by 162
Interment of monks with gloves on . . 25

PAGE

Interment of kings with gloves on . . . 33
Investiture by glove 28, 201
Irish gloves, *temp.* Hy. VII. 157
—— industry, former fame . . . 97
—— magnitude . 100
—— possibility of re-esta-
blishing 98, 102

James I., his gloves 47
Jessamine gloves 163
Jewelled gloves 18, 103
—— of kings 28
Jews, gloves among the 8
John, King, effigy of, bears gloves . . . 35
Jonson, Ben, or Dr. Johnson? 96
Judges, gloves of 54
—— forbidden to wear gloves . . . 54

Kid-rearing should be followed 182
Kid skins 181
—— monopoly in proposed 182
Kiss, gloves paid for 234
Kings, the gloves of 29, 37
Knit gloves 112
Knitting, introduction of 113
Knitter, glove, S. Anne supposed to be . 136
Knotted fringes on gloves for cyphers . . 130

Lace-trimmed gloves . . . 55, 91, 130
Ladies late in adopting gloves 106
Largess 232
Licence imposed on glove retailers . . . 173
Limerick, gloves made at 97
Linen gloves 16, 38, 129
—— symbolic 16
—— for the Child Bishop . . . 24
Love, courts of 253

Maiden assizes, gloves at 51
Maidens, gloves hung above the tombs of 240
Martial gloves 92
Markets, gloves to be sold only in . . . 154
Mary, Queen, gloves given to 41
—— of Scots, her gloves . 44, 46
Mechanism employed in glove making . 179
Miracle plays, glovers' part in 142
Miracles worked through gloves 19
Mittens 105
Monasteries, handicrafts taught in . . . 15
—— gloves made in . . . 16
Morice-dancing dress of Perth glovers . 139
Motto of glovers' company 139

	PAGE
Mourning, pomp of	245
Myrtle-scented gloves	94
Neroli gloves	92
Nettle fibre, gloves made from.	185
New year's day, popular festival	22
———— gloves given on	228
Oath, glove removed before taking	56
Ordinances of the glovers	149
Otter-skin, gloves of, recommended by Izaak Walton	103
Outlawry, reversal of, gloves given on	53
Oxford gloves	115, 248
Pedlars, gloves carried by	128
Penalty on illicit import of gloves	162, 166
Penitents of love	254
Perfumed gloves	81, 97, 163, 184
Perfuming gloves, processes employed in	89
Persians condemned for wearing gloves	9
Perth, glover incorporation of.	136, 161
———— extracts from minute book of	150
Pie-poudre court	193
Plate, gloves of.	72
Pledges, gloves as.	189
Poisoned gloves	196
Prayers, invocatory, said over gloves	17
Priests receiving presents of gloves	239
Protection, gloves the pledge of	194
Proverbs, gloves in	27
Purse and glove combined	132
Pursers and leathersellers united with glovers	149
Queen's tobacco pipe	168
Quintain, glove pledge in playing at	80
Quit rents, gloves paid for	200
Regalia, English, gloves in	31, 37
Ribbon-trimmed gloves	128
Richard I., effigy of, bears gloves	35
———— recognised by his gloves	36
Rings worn over gloves	34
Romance, XIVth. cent., gloves conspicuous in.	107
Romans, gloves of.	9
Russia leather, gloves made from	94
Ruth, book of, disputed passage in	6
Saint Anne, knitter of gloves	136

	PAGE
Sanitation, gloves in	46
Satin gloves	184
Scale-work gauntlets	71
Scott's *Fair Maid of Perth*	137, 141
Sepulture of kings, manner of	33
Servants receive and give gloves	228
Sewing gloves, remuneration for	150, 151, 152
Shakespeare, gloves of	122
———— son of a glover	158
———— tradition respecting	260
Sheepskin gloves	103
Shoe or glove ?	6
Shooting gloves	64
Signs of glovers	161
Silk gloves	21, 114
Sleeves worn as favours	258
Smuggling of gloves	163, 176
Spanish gloves	83
Spider silk, gloves made from	185
Sports, gloves as prizes in	80
Stables, not to be entered with gloves on	233
Stow, John, his account of perfumed gloves	82
St. Paul's building fund, contributions to	26
Sunday trading forbidden	149
Swearing "by the glove"	196
Sweet bagges	82, 84
Symbolism of gloves	16, 190
Taffety gloves	184
Talmud	9
Tawyers first associated with glovers	135, 158
Tax imposed on gloves	171
—— glove, proceedings in Parliament	173
Tenure held by gloves	30, 197, 200
Thebes, mural painting at, gloves on	9
Thread gloves	128
Tournament, gloves prize in	80
Universities, gifts of gloves by	246
Velvet gloves	184
Venice, luxurious excess in	93
Violet-scented gloves.	85
Wager of battel	210
Wages of glovers regulated	145
Walnut-shells, fine gloves enclosed in	97
Walpole, Horace, wears gloves of James I.	49
Walrus hide, gloves made from	184
Wedding gloves	235

INTRODUCTORY.

———◈———

"WHAT can there be in Gloves to make a book about?" This question has often been provoked during the preparation of this little volume, and, to the majority of people, will appear reasonable enough. Students of costume, and that happily increasing army of readers interested in social antiquities, will feel no such wonderment. They will know that gloves in past times have been more than mere articles of utility or convenience, and be possessed of some information upon the many points of law and usage in which gloves have been given so conspicuous a part. Still, it is hoped that most of these seekers after truth will find in the following pages many facts, traditions, and illustrations new to their knowledge, and that all will hold it a benefit to have these matters of history ready for reference.

With the general public the interrogatory would be most

properly answered, Quaker fashion, by asking another—
What is there not in Gloves to write about? It is no dry-
as-dust subject; very few figures, and no technicalities of
manufacture, will be given place in this treatise. Our
object is to narrate the numerous customs and practices in
what we may call glove ritual, and, tracing the past history
of gloves with some topics or incidents connected with
them, to make up what we may call a volume of glove
lore. Gloves have in these days no symbolism or associations. There yet lingers a practice of giving and wearing
gloves at funerals, chiefly among those who have generally

Leathern hands,
Freestone-colour'd hands—

so that, with Rosalind, we might think their old gloves
were on. But few know anything of the antiquity or institution of this unwritten law which requires this observance.
A newspaper reader, too, occasionally comes across an
account of white gloves being presented to judges who
preside at maiden assizes—another custom which most
people would find it hard to account for, or to give any
account of. With a vitality that defies Time, and is proof
against the utilitarianism with which this age is so often
twitted, these remains of the past bring down to us pre-
scriptions which doubtless had once practical force. These
two points of usage, and those now empty and without
significance, are all that remain to us of the ceremonial of
gloves, which was so elaborate and general that every

grade of society was once in some measure under its rule. Gloves protected the business of the merchant and pedlar, and were a general pledge of security ; they conveyed defiance from one knight to another, and breathed " threatenings and slaughter," even while they were the chosen token of faith between lovers, and employed on missions of peace. Between class and class they were, on great festivals, a bond of union when caste distinctions ruled supreme, and on all the eventful occasions of life— particularly at those momentous times which fill what a lady novelist delicately calls " the hatch, catch, and snatch columns of our newspapers "—gloves have been proper to the domestic rites with which they were observed, and have been mute messengers of congratulation or consola- tion. On the bench they have denoted probity, and in the Church purity ; in the helm of a knight they have been a perpetual reminder of the love of his absent "dearling," and a constant incentive to courage—a badge, openly worn, of loyalty and constancy, which none might meddle with or scoff at. There is, indeed, ample justification for writing the History of Gloves, for none other symbol—the blessed Cross of our common faith excepted—has so entered into the feelings and affections of men, or so ruled and bound in integrity and right the transactions of life.

In the history of dress, gloves stand out distinct, unique. True, some other garments have had part and lot in material affairs, and have been made indicative of weightier matters than mere wear and tear. Jews have been compelled to wear distinctive dress that they might not, under any circumstances, be mistaken for ordinary mortals. Quakers have voluntarily chosen to make their garb a part of their belief, and evidence of their separation in creed from all others. In days, far removed from our own, when the nonpayment of just debts was held to be discreditable, bankrupts in France were condemned to wear a green cap "to prevent people from being imposed upon in any future commerce." The girdle—on which, before the introduction of dress receptacles, such personal necessaries as keys, ink-horn, dagger, purse, or books were suspended—has been made the symbol of investiture or renunciation, and was so essential in mundane affairs that the imprecation became common, "May my girdle break if I fail!" They made a recusant, in mediæval times, wear a badge embroidered with a faggot, as a gentle reminder of the death by fire, from which recantation had redeemed him; and little flimsy flags, embroidered with a cognizance, which perhaps was not altogether creditable in its origin, have often served to raise drooping spirits, and been made the rallying point in the last struggle of a fierce fray. Crests and coats-of-arms have not yet lost their value as insignia of family pride; they are the last refuge of aristocrats, and the first resort of those who would become so; and other less-dignified emblems have been made the standard of political parties, from the face-patches, to which the *Spectator* testifies, to chimney-pot hats in the Radical agitation of the first part of this century. Was not Sweden, too, once racked by the rival factions of hats and caps, which were the badges of the advocates respectively of French or Russian alliances? The *bundschuh*, in Germany, made their patched shoes the evidence of their needs and demands when they revolted, as did the *camisards*, of the Cevennes, with their coarse smocks. Is not a blacksmith's apron immortalized in the royal standard of Persia, because raised in insurrection which proved successful, and made still more safe against oblivion in the pages of Carlyle? Scotch blue bonnets, and Stuart white cockades, have signalized patriotism and fidelity; and the Phrygian cap has too often been the banner in a struggle for fictitious liberty. Not to multiply examples of hidden import in dress, have not some stuffs even been endowed with moral attributes, so that they gave grace or dignity to those by whom they were fittingly worn? "Russet yeas" and "honest kersey noes," in Shakespearian numbers, declare the sturdy independence and rugged truth of the yeomen who commonly wore those ancient stuffs, as the "taffeta phrasers" and "silken terms precise" indicated the prim courtliness of another class. But all these, in their several distinctions, must make way for Gloves; for in every several respect in which these have been the outward and visible sign of hidden things, Gloves have outweighed them.

This is not the first History of Gloves that has been attempted. Some fifty years ago—to be particular, in 1834—there was issued a little volume—now very rarely

met with—by one William Hull, Junior. A large portion
of the book is taken up with

SOME OBSERVATIONS ON THE POLICY OF THE TRADE
BETWEEN ENGLAND AND FRANCE,
AND ITS OPERATION ON THE AGRICULTURAL AND
MANUFACTURING INTERESTS,

as the title-page duly sets forth. The glove trade had, nine
years before this date, been let out of leading-strings. For
some centuries previous there had been in force a prohibition
which forbade the import of any foreign gloves whatever,
which, of course, allowed home manufacturers the privilege
of making what manner of goods they chose. No fear of
competition troubled them, and they jogged along the old
beaten trade path, in the same fashion and at the same
pace as their forefathers had for generations jogged before.
This serenity was rudely disturbed by the admission of
foreign gloves on payment of specific duties. The change
found them unprepared, and much distress ensued—distress
which was at its height when Mr. William Hull issued his
little volume to waken fresh interest in home manufactures, and demolish this dreadful doctrine of Free Trade.
McCulloch, author of the well-known *Cyclopædia of Commerce*—still a standard work—had just previously declared
that the glove trade had received benefit by being stimulated into healthy rivalry, and it is against this assertion
and its author that Mr. Hull runs a tilt. "Free Trade," he
says, " abstractedly considered, is correct, but cannot be

brought into operation practically." Of the threatened free import of foreign grain, he says that, if it is permitted, "the agricultural interests, already too much depressed, must be utterly destroyed, and the home demand for manufactures decreased in proportion." He has, as we know, proved a very poor prophet, and since the glove trade, in spite of its freedom, has not yet become extinct, the controversy—in which McCulloch was, on paper, knocked down, jumped upon, and altogether vanquished— need not be further noticed here. The historical references given in the book may be likewise dismissed, since they are taken, almost without exception, and altogether without acknowledgment, from a paper in D'Israëli's *Curiosities of Literature*, where they are admittedly borrowed "from the papers of an ingenious antiquary, from the *Present State of the Republic of Letters*." In a later work on *British Manufactures*, George Dodd, generally a painstaking and worthy student of trade progress, borrows freely, but openly, from Hull's work, doing him the injustice of writing him down "Hall." At this point the history of gloves found stoppage, saving some brief articles, generally but a few lines in length, in various encyclopædias, and longer, but still brief, notices in the valuable works on costume of Strutt, Fairholt, and Planché, as well as another necessarily short notice in the *Drapers' Dictionary*, of which the present writer claims parentage.

There has thus been only one attempt to give more than a surface view of this most interesting subject; that was made subservient to political purposes, and was, in any case, written before modern archæological research had unearthed so much of that knowledge which has made us more familiar with the habits and manners of our remote ancestors than even those who succeeded them. Of this knowledge, and the channels through which it is conveyed to the public, free use has been made in these pages, and in them fully acknowledged. Of the other various sources from which information has been drawn, it is not necessary to speak here, for specific references will be found in the text, and a list of all the works consulted would make a very lengthy catalogue. Other indebtedness than that which we so often fail to give books is due in no less degree to the many persons who have, almost without exception, replied readily and with the utmost kindness to requests for information, or furnished valuable illustrations. To Miss Frances Benson, of Malvern Wells ; to Miss Mary Mayo, of Riverdale, Dorking ; to the Rev. Walter Sneyd, M.A., F.S.A., of Keele Hall, Newcastle-under-Lyne ; to the Rev. J. Fuller Russell, B.C.L., F.S.A. ; to Colonel J. S. North, D.C.L., M.P. ; to Mr. Llewellynn Jewitt, F.S.A. ; and to the authorities of the Bodleian Library and Ashmolean Museums, I owe the original illustrations which appear in these pages. In several instances I have had gloves, the worth of which could not well be esti-mated, entrusted to my keeping. It were ungracious to a degree in me if I failed to make acknowledgment of the many facilities afforded me for study in the fine

Corporation Free Library at Nottingham, and the uniform
kindness of Mr. Briscoe, its chief. To Professor Thorold
Rogers, M.P., for permission to make use of facts in his
fine *History of Agriculture and Prices ;* to Sir Reginald
Hanson and Dr. Reginald Sharpe, for translations of invaluable City records; to Mr. Kirkwood Hewat, Lord
Provost of Perth, and to Mr. William MacLeish, Town
Clerk, for aid in tracing the ancient Glover Incorporation of
the first capital of Scotland; to Messrs. Dent, Allcroft & Co.,*
for confirmation of facts in the trade history of gloves; and to
the Director of the South Kensington Museum, for information most willingly given ; and, above all, to my old friend
Charles James Wellstood, for his vigilant and appreciative
assistance—to all these I owe very hearty thanks, as well
as to that most excellent journal, *The Queen*, in which a
sketch of this present work found ready publication.

S. WILLIAM BECK.

Bingham, Notts.

---

* Mr. W. Davy, of this firm, who was my channel of communication,
interested himself greatly to discover some traces of the old Glover Company
of Worcester, which has disappeared and left " not a wrack behind." Some
of the archives were believed to have been, many years ago, in the possession
of a Mr. Birmingham, since deceased. His eldest daughter—also since
departed—was thought likely to have inherited them, and search was then
made for her representatives or executors, but without any trace of the old
records being discovered. This fact is mentioned to show how fully and freely
I have sometimes been assisted. In the hope of meeting with some information
of value, Mr. Rofe, one of Messrs. Dent's cutters, voluntarily searched the
town records of Worcester, only to find the glovers, with some other traders,
forbidden, in 1467, to throw filth into the Severn from the bridge ; a regulation
which may be found in Toulmin Smith's *English Gilds* (*Early English Text
Society*), p. 396.

A 3

[*Most of the small illustrative cuts, and particularly those to the chapter on Gauntlets, have appeared before in Planché's* CYCLOPÆDIA OF COSTUME.]

HISTORICAL.

———•◦•———

"OUT OF OLD BOKES IN GOOD FAITH
COMETH ALL THIS NEW SCIENCE THAT MEN LERE."

—Chaucer.

GLOVES.

CHAPTER I

Etymology of Gloves.

LANGUAGE is the life-blood of a nation, as much our inheritance as our kinship, as close a tie of union and patriotism as our common birth. How, in a strange land where it seems so wrong that even little children can chatter in an unknown tongue, the heart goes out to the stranger with whom we can converse! How, even among our own people, the floodgates of memory will be opened by the utterance of a rough provincialism, which speaks at once of youth and home! Language is the endowment to which conquered nations most tenaciously cling, the most invincible barrier to complete subjugation. It is part of the people. It has grown with their growth, gathered power and meaning through all the ages, until, like the thread on which a chemist clusters a mass of crystals, many a word carries with it a glittering generalization, conveying a volume of meaning. A good dictionary is truly very interesting reading, in spite of the man who declared that such an one changed the subject too often.

In many a word is tersely recorded some part of our national history, or evidence of our communion with other nations. Through this absorbent property of language it comes that, in our intellectual coinage, many pieces are included not properly current, but having great value in the exchange of speech.

It is well for us that we are beginning to regard history as made up of more than battles and sieges, revolutions and treaties. That truer history of the people is being written, in which the chronicler comes down from the mountain top, whither he had gone to watch the course of campaigns, and mentally mixes with the people, studies their feelings and faith, shares with them their struggles for liberty, rejoices with them not only in their military achievements, but also in their peaceful triumphs over the elements or matter, in their progress in the arts, sciences, and manufactures. Philology has direct bearing on such study. "How often," writes Sir George Birdwood, "the fairy-tales of trade and commerce lie hid in our most heedlessly-uttered household words;" and in our trade terms can often be traced the course of business, the origin of fabrics, either in the material, or manner, or place of manufacture, or the countries from which articles are imported, just as our progress in production may similarly be followed in our surnames. So, professedly writing a chapter of trade and social history, we first have a turn at word-hunting, which is very fine sport and a most profitable pursuit, and consider the etymology of gloves.

It cannot be pretended that the root of·the word is found beyond dispute, or its origin authoritatively settled. Philology never rests, and knows not finality. Dictionaries may some day be as secure from the assaults of change and decay as the Pyramids—but not before the Millennium. Until then one dictionary will be only published, like an encyclopædia, to be superseded by another, and until then philology, reaching forward and backward, will constantly accumulate fresh treasures, new phrases, while continually analyzing those newly won or already in firm possession. So long as genius exists, or so long as research discounts the time when all things hidden shall be made known, so long will the study of language reward its votaries with fresh successes.

We are concerned now only with retrospective philology, and require to see how far the history of the word aids its material history. The farther back the term can be dated, so much longer must gloves have been in existence. It is an indubitable fact that gloves were in use before they were so called, as will presently be shown, although we cannot doubt that the article and name came together into use among us. This, unfortunately, disposes of some very pretty and poetical theories, which would associate even the primary title with the virtues which gloves have always symbolised, or the customs with which they have ever been associated. For instance, Minsheu, one of the pioneers of " lexicographical literature," in his *Guide to the Tongues*, published in 1617, found the root of gloves in the Belgic *gheloove*, faithfulness, because gloves were the testimony of faith, and suggests an alternative derivation in " gift-love," since gloves were so often the gifts of lovers and pledges of affection. Another old writer thought the word might come from the old English *gol*, the hand, and the Teutonic *Dô* or the Anglo-Saxon *ober*, over, because gloves were put over or upon the hand. Skinner, in his *Linguæ Etymologicon Anglicanæ*, one of the works on which Dr. Johnson based his famous and, in many respects, entertaining dictionary, simply adopts these renderings. Junius, another of the burly doctor's authorities, gives the Anglo-Saxon equivalent *glofe*, and holds that to be deduced from the Danish *gloffure*, to cover, suggesting that the covering *gloffure* went through successive corruptions —*glosar*, *gloar*, and *glofe*, until it reached its permanent form of *glove*. Some modern etymologists trace the word to the old Norse *klauf*, or the Anglo-Saxon *cliof-an*, equivalent to our *cleft*, and believe the word to have originated in the first division, cleaving, or splitting of the hand-covering into stalls for fingers. Dr. Hensleigh Wedgwood says that "glove" is probably identical with the old provincial English *glave*, a slipper, from the same root with *glib* or *glabber*, slippery, *glase*, smooth. Dr. Mackay traces it to the Gaelic *ceil*, to cover, and *lamh*, the hand, pronounced *ceil-lav*, or *klav;* hence, by a natural transition, *glav* or *glove*. But the Gaelic has in it a proper name for gloves—*lamhainn*, certainly from *lamh*, the hand, but distinctly denoting a hand circle. The lowland Scotch rendering is *glu*, plural *gluwys*, as Wyntoun writes :—

> " Hawand thare-on of gold a crown
> And gluwys on hys handis twa."

Another suggested root is the Scottish *loof*, or Icelandic *loofoe*, the palm of the hand; and yet another, the Welsh *golof*, to cover. *Glove* in Germany means a hand-vow— more explicitly betokens belief or faith. " *Geloben*," says a writer in *Notes and Queries*, "in modern High German, is to vow, which, in the Low or Platt dialect, is contracted into *Globen*, and by the identity of *b* and *v* (understood by all philologists) *gloven*. As the Low or Platt dialect was that solely spoken before Luther translated the Bible into his own High dialect of Over or Upper Saxony, a Teutonic mediæval knight, throwing down the gauntlet as a challenge, and using the words, *Dat is min glove* (That is my belief), would only express the confidence of his opinion; but the act would become a symbol, and the symbol thence receive its name of glove."

The term *glofi* is common to Scandinavian nations, a fact which has led to our "glove" being very generally shown as originating with them; the parentage being sometimes shifted, for a change, to a kindred Icelandic equivalent, until, at last, Professor Skeat, indignantly claiming "glove" as an indigenous noun, says that the Icelandic is borrowed from England, and asks whether we are to write ourselves a nation whose language has not a single native word in it.

There is thus considerable disagreement between etymological doctors as to the real root of "gloves." The only indisputable fact remaining is that gloves must have been in use when our nation was in its infancy, and that our remotest ancestors took care to adopt them, either for the sake of warmth or other protective purposes; or, possibly, since human nature was in all likelihood much the same then as now, began wearing them for the sake of display. "Peacockery" is doubtless as old as the original Adam, and, in the darkest of dark ages, there may have been some effeminate fops, or some dress-loving dames who made gloves part of their panoply of pride.

CHAPTER II

Antiquity of Gloves.

AS to the remote antiquity of gloves there can be no doubt ; but there is, in one instance, considerable dispute as to the evidence on which this claim of venerable usage is based. The question turns upon the rendering of a passage in the Book of Ruth. Here, in the authorized version, the seventh and eighth verses of the fourth chapter read thus :

> Now this was the manner in former time in Israel concerning redeeming and concerning changing, for to confirm all things ; a man plucked off his shoe, and gave it to his neighbour: and this was a testimony in Israel. Therefore the kinsman said unto Boaz, Buy it for thee. So he drew off his shoe.

For "shoe" in these verses it is said we should read *glove*. On this matter, Mr. Hull, relying upon the authority of M. Josephs, "a Hebrew of great literary attainments and author of several learned works," advances the following theory :—The Hebrew *nangal* signifies to shut, close, or enclose. When followed by *regel*, the foot, it might mean a shoe or sandal, but when it stands by itself, as in the original of the passage quoted, it must be rendered "glove." It is further stated that the ancient and modern Rabbins agree in rendering the word from the original as "glove"; and that Joel Levy, a celebrated

German translator, gave, instead of shoe, his picturesque native term of *hand-schuh*, hand-shoe, by which gloves are known in Germany to this day.

The Targum, or Chaldaic paraphase of the Scriptures, made, perhaps, in the fourth century of the Christian era, translates the disputed phrase, *narthek yad*, the case or covering of the right hand—testimony apparently indisputable. But it is said, on the other side, that the Targumist tried to find adequate expression for the Greek *narthex* in a misleading idiom, since there was no word for glove in the language he used. It is obvious, however, that the paraphrasist accepted the term as a hand-covering, and not as a foot-covering; and it should be remembered that he lived some fifteen centuries nearer the only time when this question could be authoritatively settled. It is unquestionable, too, that the employment of the glove in the symbolical sense which the passage would bear by the substitution of terms, as a token of faith and in evidence of agreements, has a most remote antiquity; while the shoe is never associated in Holy Writ but with acts of humility and obeisance. Similar customs, in both respects, still obtain in the East.

The same difficulty occurs in Psalm cviii., where, in the ninth verse, among expressions of exultant triumph and vengeful threatenings, it is said, " Over Edom will I cast out my *shoe* "; and, again, the sanction of ancient usage, together with the commonest symbolism of the glove, would lead to an alteration of the passage, as it stands, in favour of glove. To throw a glove over Edom would accord with all precedent in conveying a challenge, or the utterance of a boastful promise of punishment: to throw a shoe would have no warlike significance whatever.

The question is as little likely to be determined as that of the identity of the Man in the Iron Mask, or that other as to the authorship of the Letters of Junius. *Prima facie* evidence is all that can be brought forward on either side, and the balance of that unsubstantial testimony is certainly against both passages as they at present stand.

There is a curious analogy between this little controversy and the commentary of no less a personage than Dr. Johnson upon the passage in Shakespeare's " King John," where Hubert sees the tailor

> Standing on slippers which his nimble haste
> Had falsely thrust upon contrary feet.

In this happy fidelity to absolute truth (since the shoes of Shakespeare's time have been proved to have been made " rights " and " lefts "), and no less happy touch of nature, Johnson only saw an indiscretion and error of judgment. He solemnly proceeds to set Shakespeare right by appending his opinion in a foot-note thus : " Shakespeare seems to have confounded the man's shoes with his gloves. He that is frighted or hurried may put his hand into the wrong glove, but either shoe will equally admit either foot. The author seems to be disturbed by the disorder which he describes." This hapless correction is quite enough to deter all but high authorities from heedlessly entering upon the glove-shoe controversy.

No one, of course, contends that gloves were in common use among the Jews. Bishop Tyndall included gloves with sandals among " the pompe of the disguising," and the " false signes " of women, but they are not comprehended in the very complete list of " bravery," given by Isaiah, of which women in their pride were to be deprived in the day of visitation ; and it is believed that gloves among the Israelites were worn solely by men of rank, and only by them on occasions of display. They probably formed part of the dress of kings, as ambassadors are represented in the mural paintings of Thebes bearing from some far country presents of gloves. They were certainly not articles of every-day wear, any more than they were matters of necessity, for the sleeves of both sexes were long and ample, and readily available for hand-coverings when defence against inclemency of the weather was needed.

From such necessity first came the ordinary employment of gloves. Xenophon, at the end of the eighth book of his *Cyropædia*, complaining of the unwonted luxury then recently adopted by the Persians, says that, not only did they have umbrellas borne over them in summer, not being content with the shade of the trees and of the rocks, but, in the winter, "it is not sufficient for them to clothe their heads, and their bodies, and their feet, but they have coverings made of hair for their hands and fingers." Casaubon, commenting on this passage, when quoted in part by Athenæus, declares that "neither the Greeks nor the Romans used coverings on the hands, which even country people wear to-day," and yet, referring to the Talmudic Lexicon, believes that they were previously in use among the Chaldeans : " *Chaldæi jam olim, ut videtur, iis usi: nam in Lexico Talmudico, ' magubh' exponitur manuum indumentum.*" The Chaldeans, as it seems, formerly used them, for, in the Talmud Dictionary, ' *magubh* ' is explained to be a covering for the hands. Gloves were, however, certainly well known among the Romans, if not to the Greeks. Pliny the younger, writing to Macer an account of his uncle's journey to Vesuvius, says that the amanuensis, who

accompanied the expedition with a book and all the implements of writing, "wore gloves upon his hands in winter,
lest the severity of the weather should make him lose any
time" (PLINY : *Epist.* iii. 5). Strutt is of opinion that
gloves first came to be generally adopted to protect the hands
of labouring people when working among thorns. Homer,
in the *Odyssey* (Book xxiv. 229), has described Laërtes, the
father of Ulysses, in his retirement, "while gloves secured
his hands* to shield them from the thorns," but without
affording us any clue to their shape, or the material of which
they were made. Under the emperors they were made
with fingers, and called *digitalia*, as distinct from the
chirothocæ, or glove proper, which was then made more
like a modern mitten. *Manicæ*, mentioned both by Virgil
and Cicero, were really glove sleeves, so long that they
could be brought over the hands as a muffler. *Muffulæ*
were gloves of fur for winter wear.

The only gloves shown by Hope, in his *Costume of the
Ancients*, are those of the pancratiasts—those who engaged
in a kind of gymnastic exercise, which had in it the
elements both of pugilism and wrestling. In one instance
these are mere thongs bound over, and protecting the
knuckles, and available for offence much in the same
manner as the dreaded " knuckle-duster " in the modern
days of garotting. In the only other specimen delineated
in this valuable work the outer form is the same, but

* Thus Cowper's translation. Pope renders it,
"And gloves against the thorn ; "
and Thomas Hobbes, of Malmesbury,
" Gloves of the same against the briers too ; "
but Ogilby and Chapman both give " mittens " instead of gloves. As will be
seen, this is really a distinction without a difference.

covers a kind of loose linen sheath for the arm, with scaled armour on the wrist, the whole held in place with long straps. Of this description, probably, were the gloves mentioned in the *Iliad :*

> Him great Tydides urges to contend,
> Warmed with the hopes of conquest for his friend ;
> Officious with the cincture girds him round,
> And on his wrists the gloves of death are bound.

These, however, were as much mittens as gloves. In an enumeration of the instruments of torture used in the fourth century of the Christian era are included "the gloves" (SMITH : *Rom. Antiq.*).

It would appear that some Roman epicures had then already adopted the pet theory of Mr. Ashby Sterry, that fruit, to be really enjoyed, should, so far from being cut with knives, not even be handled ; for Varro, in the second book of his treatise, *De Re Rustica*, says that olives gathered by the naked hand are preferable to those pulled with the gloves on. Athenæus speaks of a celebrated glutton who always came to table habited in gloves, so that he might be able to handle and eat the meat while hot, and so devour more than the rest of the company. These authorities, says D'Israeli, show that the ancients were not strangers to the use of gloves, though their use was not yet common. In a hot climate, to wear gloves implies a considerable degree of effeminacy. We can more clearly trace the early use of gloves in northern than in southern nations. When the ancient severity of manners declined, the use of gloves prevailed among the Romans, but not without some opposition from the philosophers. Musonius, a moralist who lived at the close of the first century of Christianity, among other invectives against the corruptions of the age, says, "It is shameful that persons in perfect health should clothe their hands and feet with soft and hairy coverings." Denunciation, however, had as little effect in stemming the tide of fashion among the Romans as in our own day. Custom and comfort combining to recommend gloves, their use soon became general. It is, of course, quite within the range of probability that the Romans introduced the wearing of gloves into this country, and that the Britons may have been prompt to recognize the comfort of the hand-clothing of their conquerors. It is no great tax upon credulity to imagine the Britons generally wearing gloves, even though it would be impossible to prove it. We know that they wore shoes of untanned leather, and we may assume that they were as likely to protect one extremity as the other.

More remote ancestors certainly anticipated the adage which declares necessity to have been the mother of invention. The cave men wore gloves. As to the time in which the cave men lived, those best able to judge are unable to decide. An eminent geologist informs the writer that it is a fair inference that they lived in the South of France before the cold of the glacial period. It is equally impossible to fix accurately the date of the glacial period. Dr. Croll and many other geologists believe that it coincided with, though not entirely caused by, an extreme excentricity of the earth's orbit. IF so, it would seem to have commenced about 240,000 years ago, to have lasted about 160,000 years, and to have terminated about 80,000 years since.

In spite of the big IF, here is an ancestry for gloves which must satisfy their most ardent admirers, for the fact that gloves were worn by these pre-historic men has been proved by Professor Boyd Dawkins, who has made this subject his own. Among the remains which recent research has brought to light in France, Belgium, and Switzerland, there were some which justified the Professor in the belief that the cave men wore gloves, not only of ordinary size, but reaching even to the elbows, anticipitating by untold ages the multi-button gloves of the Victorian era. Verily, and of a truth, there is nothing new under the sun; although the most implicit acceptance of this truism would hardly have believed that either extreme of civilisation would thus find meeting. These gloves of the cave men were, in all probability, formed from roughly-dressed skins, and sewn with elaborate needles of bone. On a perforated canine, found in Duruthy cave, was rudely drawn an indubitable glove, as may be seen in an engraving in Boyd Dawkins' work on *Early Man in Britain.*

There is no reason to doubt that the Anglo-Saxons wore gloves. They are mentioned in the poem of Beowulf, a fragment of genuine old English romance, probably written in the seventh century; and in the laws of Æthelred the Second, surnamed the Unready, made for the regulating of commerce, five pairs of gloves formed part of the duty paid to that prince by certain German merchants, called "the Emperor's men," together with two measures of vinegar at Christmas and Easter, ten pounds of pepper, and cloths —two grey and one brown. From a MS. of the tenth century Planché has engraved, in his *History of British Costume* (Knight's Library of Entertaining Knowledge), the figure of an Anglo-Saxon lady wearing a kind of muffler, covering the hand like a glove, but having only a separate division for the thumb. In the original they are coloured blue. The instance is, however, entirely excep-

tional. Strutt, in all his laborious researches, failed to find
any indication of the practice of wearing hand-coverings
being common among Anglo-Saxon ladies, with whom,
again, the voluminous sleeve, or long mantle, was made a
double debt to pay, rendering gloves altogether superfluous.
The drawing in which these mufflers appear is verified by
another MS. of a century later (*Cotton MS.*), where a
similar pair are shown on the hands of a Norman lady,
who appears in a miniature representing the presentation
of the infant Saviour in the Temple, bearing the sacrifice
of " a pair of turtle doves, or two young pigeons." With
that happy disregard of the artistic unities which led these
early painters to dress their figures in the costume then in
vogue (painting the Magi in Anglo-Saxon garb, and
making David play on an Anglo-Saxon harp), the lady is
arrayed in the long gown and graceful head-dress characteristic of that period, and we may judge that these un-
couth gloves—about as like a pair of Jouvin's as an elephant
resembles an antelope—were no novelty with the Normans.
To add to their ugliness, these gloves " have long streamers
attached to them, and over the right-hand one is a thin
gauze or fine linen cloth, in which the doves are carried,
the end appearing to pass under the sleeve of the left arm.
The mufflers themselves are very singular, and too distinctly drawn to admit of a doubt respecting their form or
object." They may, however, have been made and used
only for particular purposes in which the hands, for the
sake of cleanliness, needed a covering; and this conjecture
is upheld by the fact that gloves were yet but seldom used,
and only by persons of high degree.

CHAPTER III

Gloves in the Church.

IN common with all our old-established handicrafts, the glove trade is under greater indebtedness to the Church than, possibly, many glovers would credit. Muscular Christianity is no new doctrine. Faith and works were once literally united in a secular sense. Before corruptions crept in, and while monastic establishments maintained the simple lines on which they had been founded, their inmates were the most skilful and industrious of artisans. Weaving, illuminating, gardening, embroidery, woodwork—these and many other occupations were practised sedulously by the holy friars. The original idea of the founders of these institutions was to bring together a company of Christians who were workers. Benedict enjoins his followers to fight valiantly against idleness, the canker of truth. "Therefore," he prescribes, "the brethren must be occupied in the labours of the hands, and again at certain times in divine study." The brethren not only practised, but taught. The monastery became as much the centre of industry as of intellect; and religion was made an active worker with commerce in furthering national interests. The efforts of the brethren often resulted in raising local manufactures to great excellence, so that they obtained more than local celebrity. To the monks at Bath, for instance, is attributed much of the fame which the stout woollen cloths of the West of England yet enjoy ; and under their active auspices, we are told, the manufacture was introduced, established, and brought to perfection. In their commercial curriculum glove-making was certainly included, as well as the dressing of leather. About the year 790, Charlemagne granted an unlimited right of hunting to the abbot and monks of Sithin, for making their gloves and girdles of the skins of the deer they killed, and for covers for their books (PEGGE : *Cur. Misc.*), thus affording another proof of the use of gloves in those early times where the past and present meet.

Gloves had, however, more than this connection with the Church, for they were given distinctive employment in its rites and services. In some instances gloves were particularly required to be put on before the consecration of the Sacrament by the priests officiating. A writer in one of the early numbers of *The Antiquary* says, moreover, that " it was always looked upon as decorous for the laity to take off their gloves in church, where ecclesiastics alone might wear them. The custom still obtains in the Church of England at the Sacrament, though it is plain it had not arisen in this connection in the first instance, since in the Roman ritual the communicant does not handle the consecrated wafer. It was, perhaps, regarded as a proof and symbol of clean hands, for to this day persons sworn in our law courts are compelled to remove their gloves."

The sacred symbolism of the Church, striving to embody principles and teach truth by sight, has not forgotten to endow gloves with hidden significance. Bruno, Bishop of Segni, says that they were made of linen to denote that the hands they covered. should be chaste, clean, and free from all impurity. Durandus, Bishop of Mende in 1287, quotes authors to prove that the *chirothecæ* were worn white, and remarks : " It was specified that by these gloves the hands would be preserved white, chaste, clean during work, and free from every stain." The gloves on the hands, of Boniface the Eighth at the time of his interment were, according to Bzovius, of white silk, beautifully worked with the needle, and ornamented with a rich border studded with pearls (PUGIN : *Eccl. Gloss.*) At some subsequent period, of indeterminate date, the gloves were not invariably white, but were changed in colour, like the other vestments, according to the current Church seasons.

Gloves for bishops are an institution yet, and are advertised in the lists of clerical necessaries. They are frequently heavily fringed with gold, and always of a light lavender tint—the bishops' own colour. In the old ritual proper to the consecration of bishops a blessing was invoked on the gloves with which they were invested. This was not exceptional with gloves, for it was an ancient practice to offer invocatory prayers over the vestments with which neophytes were robed. The *Ordo Romanus*, treating of the induction of a bishop, and the Salisbury Pontifical, order the following prayer to be said when the bishop elect puts on the *chirothecæ* :—" *Immensam clementiam tuam rogamus Omnipotens et piisime Deus ut manus istius famuli tui patris nostri sicut externus obducuntur manicis istis sic internis purgentur rore tuæ benedictionis*," which may be freely translated, " We beseech Thy great mercy, O Omnipotent and Most Holy God, that as the hands of this Thy servant, our Father, are placed with these gloves externally, so internally we may be cleansed with the dew of Thy blessing." So, in another missal, we have the following

c

invocation :—"May the careful guardianship of Christ keep our hands, that they may be able to preserve the memorials of our salvation." Again, in an ancient missal of Illyricum, ascribed to the seventh century, the officiating bishop, previous to performing Mass, put on his gloves, with the prayer: "O Creator of all creatures, grant me, unworthiest of Thy servants, to put on the clothing of justice and joy, that I may be found with pure hands in Thy sight."

Pugin, giving these examples of the ancient glory of gloves, also quotes the *Monasticon Anglicanum* to prove that they were formerly embroidered with jewels; and St. Charles Borromeo says, "They should be woven throughout, and adorned with a golden circle on the outside." This circle, though of red silk, appears on the backs of the gloves formerly worn by William of Wykeham, still carefully preserved at New College, Oxford, and surrounds the sacred monogram. Gloves, lavishly decorated, frequently appear, during the Middle Ages, in inventories of church furniture. At St. Paul's, after a mitre "seeded with pearls, the gift of Bishop Richard," there are shown "also two gloves of like workmanship, the gift of the same, on which many stones are wanting." "Also two pair of gloves, ornamented with silver plates, gilt, and set with stones." Such gloves were

THE GLOVES OF WILLIAM OF WYKEHAM.

valuable enough to be left as legacies. Bishop Snell, who died in the year 1416, is, in *The Antiquities of Irishtown and Kilkenny*, recorded to have bestowed, by will, upon his cathedral " some rich presents, as gloves, pontifical sandals, a mitre adorned with precious stones," &c. At Canterbury there were : " *Cirotecæ R. de Winchelese cum perlis et gemmis in plata quadrata.*" Item: " *Par unum cum tasselis argenteis et parvis lapidibus.*" Item : " *Quatuor paria cum tasselis argenteis.*" Item : " *Par unum de lino cum tasselis et perlis* " (DORT : *Hist. of Canterbury*).

The jewelled gloves of St. Martialis are said to have miraculously rebuked an act of sacrilege, " *Ornamenta gemmarum in lucem coram testibus vomuerunt* " (They poured forth precious stones, in the light, in the presence of witnesses). Again, in Butler's *Legends of Saints*, a pair of gloves is believed to have testified to the self-denial of St. Gudula, who died in 712. When at her prayers, barefooted, a compassionate monk placed his gloves under her feet, but she rejected the proffered comfort, when, to reward her voluntary sacrifice, the gloves were taken up, and remained supernaturally suspended in mid-air for the space of an hour.

Splendid examples of the embellished gloves worn by clerical dignitaries, and peculiar to their position, occur on monumental effigies. One appears on the tomb of Bishop Goldwell, at Norwich, and another on the effigy of John de Sheppey, Bishop of Rochester, who died in 1360. The gloves worn by prelates were not, however, even in these days of extravagant display and unseemly luxury, invariably of great value. A notable contrast appears in the gloves of two bishops—Richard de Gravesend, Bishop of London, and Bishop Button, of Exeter, who

died in 1303 and 1310 respectively. The gloves of Bishop
Gravesend, worked with gold and enamelled, were priced
at five pounds (a great sum in those days); but the thicker
ones of Bishop Button only fetched tenpence the pair.
The love of dress which at this period was sapping the
energies of the Church, and making her chief sons a bye-
word and a reproach among the people, led to the passing
of sumptuary restrictions, in one of which coloured gloves
were forbidden to be ordinarily worn by the clergy,
" *rubris seu viridibus seu virgatis ;* " they might not wear
gloves red, or green, or striped ; and, in the *Ancren Riwle*
or *Regulae Inculsarum*, of the thirteenth century, persons
devoted to God are advised to have "neither ring, nor
brooch, nor ornamented girdle, nor gloves, nor any such
thing that is not proper for you to have." Long before,
in the reign of Louis le Débonnaire, the Church of Aix,
on complaint from the bishops that inferior clericals were
infringing their prerogative of wearing rich gloves, ordered
that monks were not to wear gloves of deerskin, but were
to content themselves with sheepskin ; and gloves were
thought so essential a part of the episcopal habit, "that
some abbots presuming to wear them, the Council of
Poitiers interposed in the affair and forbade them the use,
on the same principle as the ring and sandals, these being
peculiar to bishops, who frequently wore them richly
adorned with jewels." It is remarkable that, with these
ecclesiastical exceptions, gloves have no mention in the
innumerable sumptuary laws by which, from the Roman
era down to the time of the Stuarts, restrictions were from
time to time placed upon turbulent pride.

A pair of gloves are included among the bequests of
Bishop Riculfus, who died A.D. 915 ; and, for another gene-

ration in their genealogy, we have an anecdote, related by
Ordericus Vitalis, of the escape of the infamous Ranulph
Flambard, Bishop of Durham, from the Tower. This took
place in the reign of Henry I., and the historian says that,
in sliding down a rope, and " having forgotten his gloves,"
his hands were flayed to the bone. At this time all the
higher ranks of clergy wore gloves, and were buried in
them. Fitz Stephen, a monk of Canterbury, mentions
them as forming part of the pontifical habit of Thomas à
Becket at the time of his interment ; and on some investi-
gation being made in 1854, in the cathedral church of
Ross, near Fortrose, the body of one of the early bishops
was found clothed in a silk tunic and silk stockings and
gloves, the instance being remarkable as the earliest known
employment of this material in glove-making. At the
baptism of the mother of Becket, delineated in an ancient
illumination (*Royal MSS.* 2 *B.* vii.), one of the bishops
present wears a pair of long gloves, and at a later date
one of the archbishops officiating at the coronation of
Henry IV. appears in the illustrations to a contemporary
manuscript wearing white gloves.

We meet with curious confirmation of the wearing of
gloves by prelates in the accounts of a fifteenth-century
child-bishop. It was part of the coarse humour of our
forefathers to turn the established order of things topsy-
turvy on certain festivals, and give, for a time, circumstantial actuality to the common Eastern tales and as common
poetic legends, which place a peasant on a throne and send
a monarch begging. At Christmas all things were given
over to a Lord of Misrule. Over all great houses, over the
Universities, over the Inns of Court, even over the royal
household itself, did this mock prince hold sovereign sway,

with the potency of a harlequin in pantomime and the humour of a clown. His reign was a reign of laughter, provoked by any means, but principally by making men irrational children, and showing himself the greatest child of all. On New Year's Day was held the Feast of Unreason, or Feast of Fools; over which was elected a titular archbishop, or abbot, or pope. He, with a complete mimicry of the sacred office, was clad in full ecclesiastical attire, and his followers as acolytes or priests. Thus they entered the church, and played all manner of indecent pranks. They sang filthy songs and danced, ate, drank, and played dice in the church, and often on the very altar. Nothing was sacred to them. To make the occasion more subversive of regular order they would sometimes appear, not in robes usually employed, but in "fantastic disguisements," in laughter-moving imitations of animals, or in the motley of fools, and in these would perform the functions of the priesthood, saying Mass, preaching sermons, and dismissing the congregation with benedictions. These hideous parodies of sacred services, although not unknown here, were more general in France; what was permitted of religious burlesque was almost entirely confined to the annual institution of the child-bishop, a child of tender age, who was installed for the time being on the episcopal throne. He was generally every inch a bishop, performing services in every particular, and exacting homage. Wherever he went he was made much of, and sums of money, sometimes of considerable amount, were given to him. It is only just to these little prelates to admit that they seem to have had no small idea of the importance of their office, and to have often filled it with much decorum and befitting gravity.

From the proclamation which prohibited, in 1542, the procession and ceremonials over which the boy-bishop presided, may best be gathered the extent and character of this strange undertaking. Its concluding clause says that "heretofore dyvers and many superstitious and chyldysh observances have been used, and yet to this day are observed and kept in many and sundry places of this realm upon St. Nicholas', St. Catherine's, St. Clement's, and Holy Innocents' and such-like holydaies ; children be strangelie decked and apparayled to counterfeit priests, bishops, and women, and so ledde, with songes and dances, from house to house, blessing the people and gathering of money, and boyes do singe masse and preache in the pulpits, with such other unfittinge and inconvenient usages, which tend rather to derysyon than enie true glory to God or honor of His sayntes." The boy-bishop usually was elected from cathedral choirs—at first by seniority, but afterwards on account of a comely presence—on the eve of the Feast of St. Nicholas, the patron saint of children, and performed service from that day, the 6th of December, to Innocents' or Childermas Day, December 28th. He was completely apparelled in episcopal garments, and his mitre and crosier were, as we know from old church inventories, carefully preserved from year to year. Among the spoils of the dissolution of the monasteries in the reign of Henry VIII. there were, in the priory of St. Mary of Austin Canons, "a small lytell cope for a chylde bysshop," valued at 14d. ; and at St. Katherine's, near the Tower, "Sancte Nicholas cope" was priced at 12d. (*Archæologia*, vol. 43). During his time of office he, in some instances, made a pastoral visitation of the neighbouring manor houses and villages, attended by his staff of deans and prebends,

and preaching wherever he went ; the regular clergy, even
to the bishop whom he travestied, sitting under his minis-
tration and liberally rewarding the solemn jest.   In the
York *Computus* of 1396 is given a circumstantial account of
the excursion made by the boy-bishop of that year.   In
the seventh volume of the *Camden Miscellany* this is
reproduced, together with two sermons actually preached
by one of these young dignitaries.   These, it may be
remarked, by their excellence go far to excuse the occasion.
Before setting out from the city sundry purchases were
made for the child-bishop, to which prices are affixed.
He required a wax torch, weighing twelve pounds, and this
cost 4s. 3d.   A cap was bought for him, price 9d. ; a pair
of sleeves, 3d. ; a pair of knives, 14d., which were perhaps
not so proper to his episcopal habit as to his boyish long-
ings ; a pair of spurs, 5d. ; lamb's wool for his overcoat,
2s. 6d. ; and furs to keep him warm, 6s.   For the making
of a gown for him 1s. 6d. was paid.   They bought him, too,
a pair of linen gloves, 3d., and for his attendants twenty-
eight pairs, at a cost of 3s. 4d., which, if we take this to
be an accurate copy of established order—as we fairly may
—would show that in religious ceremonies gloves were
then worn by the inferior clergy and attendants.   This
certainly was true at a later period, for Du Vert, a seventeenth-century writer, says that priors wore them in his
time, and also that cantors in many churches did,
when holding staves, wear gloves.   White gloves were,
according to the same author, used in processions, and
in some places by those who bore the reliquaries ; while
it was the custom in the monasteries of Clugni to
inter monks with their gloves on.   Gloves, declares Pugin,
" may be worn with propriety by all who in ecclesias-

tical functions carry staves, canopies, reliquaries, candle-
sticks, &c."

Doubtless many a splendid glove, hallowed by long
association, if not by direct consecration, disappeared during
the Reformation, when zeal was inspired by plunder, and
the work of spoliation went so merrily on that, as old
Fuller tells us, men " hung their halls with altar cloths, and
used copes for coverlets." In the inventory of " the plate,
juelles, ornaments, vestments, copes and bells of the Cathedral Church of the Blessed Trinitye, in Winchester," made
on the 23rd of October, 1552, by the commissioners who
did the bidding of the new Defender of the Faith, there
occurs, " j payre of red gloves with tasselles wrought with
venis [Venice] gold " ; and at the Cathedral of Durham
there was " j myter with a payre of pontyfycalls," which
are believed to have been gloves (*Archæologia*, vol. 43).
The use of gloves appears to have been maintained even
after the radical change in the Church, although, except as
bishops for obvious purposes employ them in confirmations,
they have not at present part or lot in church services.
James Montagu, Bishop of Winchester, who died in 1618,
left in his will " unto my good and diligent seruant Nicholas
Younge . . . one 200 markes in monny, and all my scarlet
gownes and robes, and rich gloues," valued in the aggregate
at £33 6s. 8d. (*Archæologia*, vol. 44). The connection of
gloves with the Episcopate, moreover, appears to have
been maintained after the Reformation in the manner least
agreeable to the bishops ; for, from an " Order in Council,"
given at the Court of Whitehall, the 23rd day of October,
1678, it appears that a custom was still enforced by which
it had been usual "to make presents of gloves to all persons that came to the Consecration Dinners and others,

which amounted to a great Sum of Money, and was an unnecessary burden to them." At this time money was required to rebuild St. Paul's. Constant appeals made for contributions towards the "reparation" of that building had been summarily stopped by the Great Fire, and a more urgent need created. Charles, with that ready good-nature which allowed him to do a good action gracefully at somebody else's expense, was now quick to perceive the "unnecessary burden" imposed by gloves on the bishops-elect, and at this Council, " taking the same into his consideracion, was thereupon pleased to order that for the future there shall be no such distribucion of gloves, but that in lieu thereof each Lord Bishop, before his Consecracion, shall hereafter pay the sum of 50*l.* to be employ'd towards the Rebuilding of the Cathedral Church of St. Paul's. And it was further ordered that His Grace the Lord Archbishop of Canterbury do not proceed to consecrate any Bishop before he hath paid the said sum of 50*l.* for the use aforesaid, and produced a receipt for the same from the Treasurer of the Money for Rebuilding the said Church for the time being, which, as it is a pious work, so will it be some ease to the respective Bishops in regard the Expense of Gloves did usually farr exceed that sum."

The contributors to the fund for rebuilding St. Paul's Church included the following bishops, who paid in lieu of dinners and consecration gloves :—

June 30, 1668. Dr. Edward Raynbow	Ld. Bp. of	Carlisle	...	£50 0 0
Jan. 7, 1669. Dr. Walter Blandford	,,	Oxford	...	50 0 0
Mar. 28, 1672. Dr. Nathanael Crewe.	,,	Oxford	...	50 0 0
Oct. 30, — Dr. Thomas Wood...	,,	Coventry and Lichfield		50 0 0
Ap: 26, 1673. Dr. Peter Gunning ...	,,	Chichester ...		50 0 0
May 17, 1678. Dr. Hy. Compton ...	,,	London	...	50 0 0

Feb. 10, 1679.	Dr. W. Gulston ...	Ld. Bp. of	Bristol(G.&D.)£100 0 0
June 19, 1679.	Dr. William Bean...	,,	Landaffe ..	100 0 0
Dec. 6, 1680.	Dr. Robt. Frampton	,,	Gloucester ...	100 0 0
Sept. 4, 1683.	Dr. John Fell ...	,,	Oxford ...	50 0 0
June 3, 1684.	Dr. Thos. Spratt ...	,,	Rochester ...	200 0 0
—	Dr. Wormock ...	,,	St. David's ...	100 0 0
July 22, —	Dr. Thos. Smith ...	,,	Carlisle ...	100 0 0
Jan. 26, —	Dr. Thos. Ken ...	,,	Bath & Wells	100 0 0

—(DUGDALE : *History of St. Paul's.*)

The sententious wisdom of proverbs has, in several instances, made use of gloves to point a moral. " Touch not a cat but (without) a glove," is a Scottish axiom, which remains as the motto of a well-known firm of publishers. " Cats that go ratting don't wear gloves," says a Spanish proverb, with the same purpose as Mr. Smiles describes kid gloves on the hands of young engineers as "non-conductors," checking the current of education by interposing between fingers and work. These maxims are obvious enough, but a proverb is recorded by Hazlitt, which is inexplicable. " What, a bishop's wife ! eat and drink in your gloves !" What peculiar privilege this was, and why it was particularly enjoyed by bishops' wives, will probably now never be determined.

CHAPTER IV.

Gloves on the Throne.

GLOVES had in former times a recognized place in
regalia, and in this relationship are believed to
have a long descent of dignity. They were in this
country an appanage of royalty; the imperial ornaments
of Germany included "the gloves, embroidered with curious
stones;" and preserved at Vienna are "gloves of the imperial costume, decorated with enamels" (LABARTE).
Sometimes they were more particularly distinguished by
being dyed in the purple so long distinctive of royalty.
"Purple gloves, ornamented with pearls and precious
stones, were anciently deemed ensigns of imperial dignity,
as is recorded by Pachymenera and other authorities"
(HULL).

We meet with many instances proving that gloves were
intimately associated with kingly power. Monarchs were
invested with authority by the delivery of a glove. A
Register of the Parliament of Paris, dated 1294, says that
"the Earl of Flanders, by the delivery of a glove into the
hands of the king (Philip the Fair), gave him possession of
the good town of Flanders;" and Favyn observes that the
custom of blessing gloves at the coronation of the kings of
France—a ceremony which was maintained as long as

kings remained as rulers of the country—"was a remain of the Eastern practice of performing investiture." *An Historical and Chronological Treatise of the Anointing and Coronation of the Kings and Queens of France*, by M. Menin, which was "done into English" in 1723, shows the gloves to have been put on the king after the ceremony of anointing, being first blessed by the officiating archbishop saying over them a dedicatory prayer, and sprinkling them with holy water. During this preliminary, the archbishop put off his mitre, then, again donning the mitre, he put the gloves on the hands of the king, at the same time offering up another prayer. After the coronation had concluded it was the established custom for the king to be taken to the palace of the Archbishop of Rheims, in which town the ceremony was generally performed, and, in a chamber allotted to him, take off the gloves and shirt, which were given to the chief almoner to burn, "since having touched the Holy Oil they ought not to be profan'd by other Uses." As an exception to this rule, it is noted that, at the coronation of Louis XIII., Mary de Medicis, his mother, "had the Piety to desire them, in order to preserve them carefully in her Cabinet, which was granted."

In the papers quoted by D'Israeli is instanced an incident in which a glove was, under romantic circumstances, taken as the actual representative of power. Young Conraddin, the last of the Hohenstaufer male line, having fallen into the hands of Mainfroy, who had usurped the crown, was, in 1282, brought up for execution. On the scaffold the unfortunate prince made heavy lament over his cruel fate, and publicly asserted his right to the succession. In proof of this he cast his glove among the assembled crowd, entreating that it might be conveyed to his relations, who would avenge his death. It was taken up by a knight and brought to Peter, King of Aragon, who, *in virtue of the same glove,* was afterwards crowned at Palermo. So the kings of France, when at the point of death, sent or gave the imperial ornaments to their sons in token to invest them in the kingdom.

In this practice of conferring investiture by the gift, or perhaps the transfer of a particular glove, may well have arisen the close connection between gloves and royalty. From the emblem of possession to him who gave possession was but a short step. The gift of lands by the king was the very essence of the feudal system, in which modern society had its rise, and the lien of the king over all land was the first doctrine of Divine right. Thus the glove, by which tenure was granted, was the pledge of the service by which tenure was held, and, on the hands of him who could grant the one and demand the other, must have been a right royal symbol. It is noticeable that the glove in the attire of monarchs has particular prominence under the Norman and Plantagenet dynasties, when the feudal system was yet young. Gloves were then part and parcel of kingly power. The king even relegated by them his power to others. The glove was his ambassador, and as representative of pains and penalties as a policeman. Under the protection of the king's glove fairs were instituted and maintained—a part of our subject to which fuller consideration will hereafter be given.

We have a good idea of what these royal gloves were like, for they appear frequently in mediæval manuscripts, where they are always shown white, and with very wide pointed cuffs. In the engravings from those sources given in Strutt's *Regal and Ecclesiastical Antiquities,* such gloves frequently occur. An illumination of the time of Edward I. gives a representation of the assassination of Richard I. at Chalus, and the king carries in his hand a pair of white gloves. In a fine illumination prefacing a book written and embellished by Lydgate, a poet and monk of Bury St. Edmunds, and by him presented to Henry IV., that monarch, shown in the act of receiving the gift, wears a pair of gloves of this peculiarly royal shape. In another fine series, which Strutt reproduces *in extenso*, are delineated several incidents in the career of Becket, the martyr of Canterbury. One of these shows the king pronouncing sentence of banishment against all the kindred of the outlawed prelate when he fled before Henry's displeasure. The king, in a dignified attitude, stretches out a sword in his gloved hand, an attendant courtier, as though it were his office, bearing the other glove behind him.

There is no authoritative evidence of the early employment of gloves in English coronations, although there is fair presumptive proof of their having had part in the ceremony. Among the ancient regalia seized by the regicide Martin, in 1642, and destroyed seven years later, there is included " One paire of gloves embrodrd wth gould," valued at one shilling. The robes and other coronation requisites, which must have been peculiarly obnoxious to fanatic Roundheads, were always believed to have been the original insignia of Edward the Confessor, with the exception of " King Alfred's crowne," which, we are told, was "of gould wyerworke, sett with slight stones and two little bells." They were thus, as the French regalia was associated with Charlemagne, popularly connected with ·the two princes who, for valour and piety, were formerly most held in peculiar veneration by the people. Mr. Planché says, we know not why, that, whatever claims to remote antiquity the English sceptre and crowns may have had, the garments could have none. At any rate, they must have shown many signs of age. Before being "totallie broken and defaced," in accordance with an order of Parliament, an inventory of all these emblems of departed glory was made, while they were in the charge of Sir Henry Mildmay, in the upper jewel house of the Tower, whither they had been brought in direct contravention of ancient usage, which, established by the foundation charter of Westminster Abbey, granted by Edward the Confessor, and fortified by the bulls of successive popes, directed that they should, from coronation to coronation, be held in charge of the Abbot and monks of Westminster. Henry VIII., eager to mark by every means his new departure, broke through this rule when he abjured the Papal supremacy, appropriated the ancient insignia to the use of the Crown, and removed it from Westminster. He made it the heirloom of a family, rather than the sacred property of a nation, complains Taylor in his *Glory of Regality.* To the inventory of these ancient effects is happily appended an appraisement of each several article, and the prices speak either very strongly of Republican depreciation, or are very significant of their dilapidated condition. Thus, "One common taffaty robe, very old," and "One robe laced with gould lace" are valued at ten shillings each, "one robe of crimson taffety sarcenet" at five shillings, "one paire of buskins, cloth of silver, and silver stockings," again described as "very old," are priced at half-a-crown, and "one paire of shoes of cloth of gold" at two shillings.

" One silver cullerd silk robe," and " one old comb of horne," are contemptuously described as " worth nothing." There is certainly, considering the care with which this apparel would have been preserved, no manifest absurdity in believing that it might possibly be dated back to the eleventh century. There is, besides, no manufacturing anomaly in assigning that date to them, for silks of great splendour were known and worn in those early times, and the skill of Anglo-Saxon ladies in " painting with the needle " had given their work so great a reputation for excellence that *Anglicum opus*, English work, was celebrated all over the Continent.

In the accounts of several coronations given by the early chroniclers—Langtoft, Hoveden, Matthew Paris, and others —even though some give particulars of the ceremonial observances, no mention is made of gloves being used ; but, again, against this negative evidence we may set the fact that kings were buried with gloves on their hands when, arrayed in ghastly state, they were gathered to their fathers. Very much after the manner of their cousins of Egypt, and in accordance with a practice common to all classes from Anglo-Saxon times, kings were carried to the grave habited in every particular in the garments which had in lifetime marked their high estate. They even laid beside Charlemagne, in his sepulchre, the gilt travelling pouch which he used to wear when he went to Rome, much in the same manner, and perhaps with the same motive, as the fee for Charon was often put in the hand or mouth of a corpse that the passage over the Styx to the Elysian fields might not be checked or delayed.

Among this funereal pomp of our monarchs gloves were carefully included. In the " auncyent manner of the sepulture of Kings in this realme," set forth in Hearne's *Collection of Curious Discourses*, it is said, " The corps, preciously embalmed, hath been apparelled in royal robes or estate, a crowne and diadeame of pure goulde put uppon his head. Having gloves on his hands, holding a septer and ball, with rings on his fingers, a coller of gould and precious stones round his neck, and the body girt with a sword, with sandalles on his leggs, and with spurrs of gould."

The burial of Henry II. is the first of which we have particulars. Matthew Paris and Simon of Durham both agree in stating that he was borne to the grave arrayed in royal robes, crowned, and bearing the sceptre, with his sword by his side, with shoes wrought with gold on his feet, and spurs fastened on them, on his hands gloves, and a ring on his finger. One account adds that, as Henry's son Richard came to the side of the corpse, blood flowed from the nostrils. It is possible that the ring was worn over the glove, for such meretricious display has been not infrequent. Bishop Hall, in his *Toothless Satires*—which are by no means so harmless as the title would lead one to suppose—falls foul of this folly :

> Nor can good Myson weare on his left hond
> A signet-ring of Bristol-diamond,
> But he must cut his glove to shew his pride,
> That his trim jewel might be better spy'd.

We know that gloves were subsequently commonly put on the bodies of people of high rank in sepulture ; and unless the rings were worn outside the glove it would be hard to account for the notices occurring of the hands being covered with them. Thus, in the account of the funeral ceremonies of Queen Elizabeth, wife to Henry VII. (*Antiq. Rep.* iv. 657), where an image made and clothed to represent the deceased was borne in the procession under a canopy, it was seen to have the "fingers well garnished with gould and precious stones." With such fidelity were these counterfeit presentments of the departed prepared that, according to this account, not only was the "image" clothed in the very robes of estate of the queen, but it had "her heire about her shoulders." At the burial of Edward IV. there was also directed to be made and furnished "an ymage like hym clothed in a surcote, with a mantell of estate, the laces goodly lying on his belly, his sceptre in his hand, and a crowne on his hed, and so cary him in a chare open, with lights and baners" (*Archæologia*, i.)

If corroboration were needed of the ancient accounts of the burial of Henry II. it would be found in the description of the monumental effigy discovered above his tomb at Fontevraud, in Normandy. Stothard, who literally unearthed this valuable monument with others of our English kings who were laid to rest there, in his *Monumental Effigies of Great Britain*, gives in detail the dress of the figure, and of the gloves says, that they have jewels on the centre of the back of the hand, a mark of royalty or high ecclesiastical rank. Richard I. and John, both buried at Fontevraud, have, like their father, richly jewelled gloves. On opening the tomb of John it was found that the body within was carefully counterfeited in the effigy above, so that these tombs may be taken as literal representations of the deceased kings as they were buried. John's effigy at Worcester likewise displays the jewelled glove which was so eminently a mark of sovereign power, and in the wardrobe of his son, Henry III., were included two pairs of such gloves. *Et de ii paribus*

chirothecarum cum lapidibus (Rot. Pip. an. 53 *Hy. III.)*
According to tradition, Richard was recognized in Austria
on his return from the Crusades by his jewelled gloves.

The writer of the article on gloves in the yet incomplete
edition of the *Encyclopædia Brittanica* says that gloves
were found on the hands of Edward I. when, in the
interests of history, his tomb was opened in 1774. Other
authorities remark that the practice of placing gloves on
the hands of deceased monarchs was so commonly
followed, that it was a matter of wonderment that *no gloves*
were found on the corpse of Edward I. when disinterred.
These contrary assertions were very embarrassing; something was palpably wrong somewhere. Statements so
diametrically differing are rarely heard outside of a police-
court. However, the matter is more easily settled than
when two quarrelsome neighbours, each accuse the other
of a first assault; for in the *Archæologia* (iii. 376); as well as
in the *Gentleman's Magazine* for 1774 (xliv. 233), complete
accounts are given of the occasion on which several gentlemen of the Society of Antiquaries were allowed to open
the stone sarcophagus, in which the remains of the
English Justinian were believed to be preserved. Curiosity
—or rather a spirit of enquiry—had in the first place
been aroused by the warrants issued in the reigns of
Edward III. and Henry IV. to the Treasury, to renew
the wax about the corpse—*de cera renovanda circa corpus
Edwardi primi*—which led to the belief that more than
ordinary care had been taken to preserve this renowned
King. The body was found, when the tomb was opened,
in perfect preservation, wrapped in two outer coverings,
one of them of gold tissue strongly waxed and fresh.
" The body was habited in a rich mantle of purple,

diapered with white, and adorned with ornaments of gilt
metal, studded with red and blue stones, and pearls. The
mantle was fastened on the right shoulder by a magnificent
fibula of the same metal, with the same stones and pearls.
His face had over it a silken covering, so fine and so
closely fitted to it as to preserve the features entire.
Round his temples was a gilt coronet of *fleurs-de-lis*. In
his hands, which were also entire, were two scepters of gilt
metal, that in the right surmounted by a cross fleuri, that
in the left by three clusters of oak leaves, and a dove on
a globe; this scepter was about five feet long. The
feet were enveloped in the mantle and other coverings,
but sound, and the toes distinct. The whole length of
the corpse was 5 ft. 2 in." Sir Joseph Ayloffe, to whom
the commission was granted, and under whom the
proceedings were conducted with all decorum, in his
account of them says—"*There did not remain any
appearance of gloves*, but on the back of each hand, and
just below the knuckle of the middle finger, lies a
quartrefoil of the same metal as those on the stole, and
like them ornamented with five pieces of transparent
paste." Sir Joseph devotes further consideration to the
absence of gloves on this occasion, finding in that respect
no admissible argument for the monarch having been
entombed without those parts of established sepulchral
dress. "It hath been before observed," he remarks,
"that our Kings, when carried to their sepulchres, were
habited nearly in the same manner, and adorned with
the like regalia as at the times of their coronations, and
the ancient coronation rituals and ceremonies direct
that on those solemnities gloves shall be placed on the
King's hands; and that such gloves shall be made of

fine linen. If then, conformable to that practice, and the mode prescribed by the regulations *de exequiis regalibus,* gloves were placed in the hands of King Edward's corpse, and such gloves were made of so slight a material as fine linen, they could not long have resisted the injury of time, but necessarily must have long since perished and fallen into dust. That this was the fact in the present case is clearly evident from the quartrefoils of goldsmith's work, which, according to the regulations *de exequiis regalibus,* were to be fixed on the gloves put on the defunct, being still remaining on the back of King Edward's hands." It is perhaps far more likely that the gloves had been removed, for some cause or another, on one of the former occasions when the tomb had been opened. It is hardly feasible that so durable a material as linen should have been so utterly destroyed as to leave "not a wrack behind," when other fabrics remained in a state of comparatively perfect preservation. In a very common and analogous instance where linen is similarly employed, that is, in the wrappings of mummies, linen finer by far than that obtainable in England in the fourteenth century, has resisted the ravages of time, not for three, but thirty and three centuries. If linen gloves were placed on the hands of the first Edward when he was buried, there would, unless they were removed, certainly have remained some traces of them in 1774.

The place occupied by the gloves in the coronation ritual is, as in France, subsequent to the prayer said by the officiating archbishop after the ceremony of anointing has been performed. They are then brought forward by the Lord Chamberlain to be placed on the hands of the sovereign. To this functionary is also granted the privilege of bringing to the sovereign, on the morning of installation, the principal robes and requisites, upon which he claims all the bed and furniture of the sovereign's chamber as his fee, and for the important service of bringing water for washing the sovereign's hands before and after the coronation dinner he is allowed the basin and towel for his pains. For the coronation of James II. and his Queen, a pair of linen gloves were among the requisites ordered to be made ready, though others than these were made use of during the ceremony. An account of the ritual observed says that, after the ceremony of anointing—" The Dean of Westminster dried the places anointed, except the head and hands, with cotton wool, and again tied the ribbons that closed his garments ; a shallow coif of linen was then put on the King's head, and linen gloves were put into his hands because of the anointing, and in the meantime a short anthem was sung. Afterwards the Archbishop drew off the linen gloves, and put the ring, with a ruby, on the fourth finger, and a rich glove being presented to the King he drew it on over the ring. (BLOUNT : *Jocular Tenures : Hazlitt's Ed.)*

Monarchs, being only mortal after all, have worn gloves after the manner of ordinary folk, merely for comfort or convenience, or what not. In a *Copy of a Roll of Purchases made for the Tournament at Windsor Park,* 6 *Edw. I. (Archæologia),* there are included " half-a-dozen pair of double gloves," 35s., and "six pair of buckskin gloves for the King," 6os. A glove once belonging to Henry VI. is engraved in the third volume of the *Antiquarian Repertory,* from drawings made at Bolton Hall in 1777 ;

and, as our illustration, reproduced from that volume, will show, had not any exceptional elegance to boast of. It was made, too, of homely materials, for in the minute description which accompanies the plate we are told that it is " made of tann'd leather, lined with deerskin, with the hair on the outside, and turns down with the top ; from the end of the middle finger to the top, eight inches, top five inches, width at the thumb four inches, width at the

GLOVES OF HENRY VI.

top five inches three-fourths." In 1474, sixpence was paid for the gloves of the Scottish Queen (MAITLAND), and in 1498, four shillings for a dozen of leather gloves for the King (HENRY: *Hist.*) In the *Wardrobe Accounts of Edward IV.* are shown "viij dosen pair" of gloves without particulars ; and again, in another warrant, "gloves xviij paire." Grose quotes an entry from an account of Henry VII. in the Remembrancer's Office—

" Item, for three dozen Leder gloves, xijs." In June, 1531, the *Privy Purse Expenses of Henry VIII.* are charged with " xj payr of gloves, ijs. ixd. ; " and in the following year there is an entry of 4s. 1od. " paied the same daye to Jacson for certeyne gloves fetched by the serjeant apoticary." The *Privy Purse Expenses of the Princess Mary*, Henry's daughter, contain several interesting items relating to gloves :

1542 3.—Itm, geuen to Stephen Bonnynato geuing my lades (grace) gloves, &c. .... vij^s. vj^d.

Itm, geuen to ij children of the Chapell geuyng a payr of gloves to my lades grace .... ij^s. vj^d.

1543 4.—Itm, to godsall man bringing a payr of gloves ...... ij^s.

The *Wardrobe Accounts of Prince Henry*, son of James I., for year 1607 *(Arch.* xi.*)* contain quite a collection of gloves, among a most extensive wardrobe. So lavish, indeed, is the outlay on dress for this short-lived Prince, that it appears incredible that these accounts, which total over £4,574, could apply to one year only, even if he had inherited his mother's extravagance ten times told and been allowed full play for it. There are shown :

One pair of gloves lined through with velvett and laid with three gold laces, and gold fringe curled, 6os.

Two pair of cordevant gloves, perfumed and laid with broad silver lace and fringe curled, at 32s. the pair.

Four pair of staggs leather gloves, perfumed and fringed with gold and silver fringe, at 16s.

Six pair of plain gloves with colored tops, being very well perfumed, at 6s.

Six pair of plain gloves with colored tops and some white tops, at 3s.

Twelve pair fine gloves stitched, the fingers and the tops white silk and silver, and some trymmed with taffata and reben, at 11s.

Thirty-one pairs in one year, for a boy of fourteen! The outfit is not only elaborate, but extensive, and

GLOVES OF QUEEN ELIZABETH
(BODLEIAN LIBRARY).

includes all kinds of apparel even to goloshes and muffs, which were at this time worn by men.

Some historic gloves, traditionally associated with royalty, are still preserved in private collections and public museums. In the Bodleian Library visitors are shown a pair of gloves believed to have been worn by Elizabeth — "that virago," as Carlyle calls her, in strong contrast to the title of "bright Occidental Star," given to her Majesty by the translators of King James's Bible. Macray, in his *Annals of the Bodleian Library*, says these gloves "were worn by Queen Elizabeth when she visited the University in 1566." Elizabeth was, we know, very vain of her hands. Du Maurier *(Memoires pour servir à l'Histoire de Hollande)* writes how he had heard from his father " that, having been sent to her, at every audience he had with her Majesty, she pulled off her gloves more than a hundred times to display her hands, which, indeed, were very beautiful and very white." But either the gloves were not made to fit very closely, or Elizabeth's hands were on too large a scale to suit modern ideas of beauty in the hands of women. The middle finger of ⌐ the glove is 4¾ inches in length, and the thumb 5 inches ! The palm is 3½ inches in width ! The glove is close on half a yard long, the gold fringe at the bottom only taking two inches from the total length. They are of excellent material—a very fine white leather, worked with gold thread, edged at the bottom with yellow, and lined in ⌐ the cuff with drab silk.

· A melancholy interest attaches to a pair of gloves exhibited by their owner before the Archæological Institute in 1861—when a notable number of fine specimens of art industry were gathered together. These were averred to have been given by Charles I., on the scaffold, to William Juxon, Bishop of London, and to have been subsequently preserved by the bishop's descendants, at Little Compton, Gloucestershire. Again, it is recorded of Lady Jane Grey—the queen of thirteen days—that, when led to the block, she, after kneeling awhile, " stode up, and gave her maiden, Mistress Tilney, her gloves and handkercher." The royal victim of the following reign, Mary Queen of Scots, is believed to have left a similar memento, at present deposited in a very excellent museum at Saffron Walden. This splendid glove, which, irrespective of any other claims, undoubtedly belonged at some time to a personage of high rank, is thus described in the

Abridged Catalogue of the Saffron Walden Museum :—
" This curiously embroidered glove was presented by the
unfortunate queen, on the morning of her execution, to a
gentleman of the Dayrell
family, who was in attendance upon her at Fotherin-
gay Castle on that occasion,
February 8th, 1587. It is
the property of Francis
Dayrell, Esquire, of Camps."

GLOVE OF MARY, QUEEN OF SCOTS
(SAFFRON WALDEN MUSEUM).

From a Lithograph in the *Reliquary.*

The glove is made of a light, cool,
buff-coloured leather, the elaborate
embroidery on the gauntlet being
worked with silver wire and silk of
various colours.   The roses are of
pale and dark blue, and two shades
of very pale crimson ; the foliage
represents trees, and is composed of
two shades of æsthetic green.   A
bird in flight, with a long tail, figures
conspicuously among the work.   It
should be here mentioned that the
embroidery shown in the drawing is
repeated in *fac simile* on the other
side of the glove, and this, having
been lying against the lining of the
glass case, has retained the colour
better than the side which has
been so many years exposed to the
light.

That part of the glove which forms
the gauntlet is lined with crimson
satin (which is as fresh and bright as
the day it was made), a narrow band
being turned outwards as a binding to the gauntlet, on to which is sewn
the gold fringe or lace, on the points of which are fastened groups
of small pendant steel or silver spangles.   The opening at the side
of the gauntlet is connected by two broad bands of crimson silk, faded

now almost to a pale pink colour, and each band is decorated with pieces of tarnished silver lace on each side.*

There seems to be no doubt of the authenticity of this relic, which has been treasured for generations past by the Dayrell family. There is, further, no question but that an ancestor of the house was present at the execution of the unfortunate queen, for a copy of a letter found in the Tower Records gives an account of the execution on the day of its occurrence, and is sent by "Mar. Darell, ffrom ffatheringaie Castle, viijth of ffebruarye, 1587, to the right worshipp^{le} Mr. Willm. Darell, Esquire, hat his house at Littlecott." The writer speaks of being an eye-witness on the occasion, and describes the manner of procedure, and the bearing of the Scotch queen in enduring the fatal stroke. But it is suggested that, as every particle of dress which had been touched by the blood, with the cloth on the block and scaffold, was immediately destroyed, that no relic should be carried off with which to work imaginary miracles (FROUDE), Mr. "Mar. Darell" may have had a more important part to play on the occasion than that of a mere spectator; that, indeed, he may have acted as executioner, and received from the queen, in place of the usual headsman's fee, her glove, she being aware that a gentleman of quality had undertaken the odious duty.

Another pair of gloves, believed to have belonged to Mary, Queen of Scots, are preserved in the Ashmolean Museum,

---

* This description, with some other information, is taken from the *Reliquary*, a quarterly antiquarian magazine of long standing, edited by Llewellynn Jewitt, Esq., F.S.A., by whose permission the accompanying engraving is reduced from a full-size lithograph of the glove, illustrating the article. This lithograph, I am informed by Mr. Maynard, the curator to the museum, may be fully relied on as correct in every particular. This is the more important, because Fairholt's drawing of it (copied by Planché) is, in such case, inaccurate in its details.

Oxford. These, though not so elaborate as those just treated, are still of very fine workmanship, and, like those of Elizabeth, are of such liberal dimensions, as to tell either against the skill of the glover or the beauty of the wearer. A writer on modern dress protests against the wearing of tight gloves, because "fulness" follows on the compression of the flesh, and, "in spite of the elasticity of the lamb's skin, the suppleness of the articulation is lost." M. Blanc could hardly have found this artistic defect, or any like ground of complaint against the æstheticism of figure lines in the regal gloves of the sixteenth century, unless indeed, they had, in the other extreme, offended his sense of the "eternal fitness of things" by being unduly baggy.

The gloves worn at this period were generally of great splendour. Rank was finding in lavish display and rich ornament a refuge from the imitation of the vulgar

GLOVES OF MARY QUEEN OF SCOTS
(ASHMOLEAN MUSEUM).

which in such matters is not considered such excellent flattery. The gorgeous gloves of Mary, Queen of Scots, help us to an estimate of the splendour of those worn by her royal rival, and particularly of some presented to her by Lord North, in 1581. The following

GLOVES OF JAMES I.
(In the possession of Rev. J. Fuller Russell, B.C.L., F.S.A.)

entries occur in his lordship's *Household Book* (*Archæologia*, vol. 19) :—

> Froggs and Flies, for the Queen's gloves, 50s.
> Gloves for the Queen, 15s. ; for myself, 7s.

The " froggs and flies " were undoubtedly natural ornaments, in costly materials, for the embellishment of some fine presents for her Majesty.

Another glove, declared to have been worn by a king, was likewise brought before the Archæological Institute on the occasion before mentioned, lent by the Rev. J. Fuller Russell. They are of strong brown leather, lined with a soft white skin, the seams sewn with silk and gold thread. The embroidery, in gold and silver thread, is worked on dark crimson satin, and the cuff, lined with crimson silk, is edged with a full fringe of crimson silk and gold thread. The cuff-bands, of crimson ribbon, are edged with open-wrought loops of gold wire. They were at one time in the possession of Horace Walpole, at Strawberry Hill,* and before that had belonged to Ralph Thoresby,

---

* These gloves are made additionally interesting by a letter from Crofton Croker, containing an excerpt from one of Walpole's, and connecting them personally with that versatile and extravagant genius, in a manner, too, quite in accordance with his character. This letter is in the possession of the owner of the gloves, by whom it was, with them, kindly lent to me :—

"ROSAMOND'S BOWER, FULHAM, 1st *June*, 1842.

" DEAR MR. ANTHONY,—As you are the fortunate possessor of James the First's gloves, from Strawberry Hill, the following passage in one of Horace Walpole's letters, dated May, 1769, which I have just stumbled upon, cannot fail to interest you, and I think that it gives much additional value to those relics. Walpole writes thus :—'Strawberry has been in great glory. I have given a festino there that will almost mortgage it. Last Tuesday all France dined there ; Monsieur and Madame du Chatelet, the Duc de Liancourt, three more French ladies, whose names you will find in the enclosed papers, eight other Frenchmen, the Spanish and Portuguese Ministers, the Holdernesses, Fitzroys—

E

most zealous of antiquaries. In the catalogue of his museum (*Ducatus Leodiensis*) they are described as " A pair of King James the First's, embroidered with (?) crimson silk, and lined with the same coloured silk, the seams covered with gold edging."

in short, we were four-and-twenty. They arrived at two. At the gates of the castle I received them, dressed in the cravat of Gibbons' carving, *and a pair of gloves embroidered up to the elbows, that had belonged to James I.* The French servants stared, and firmly believed this was the dress of English country gentlemen.'—Very truly, dear Mr. Anthony, yours,

" T. CROFTON CROKER."

This cravat was carved by Grinling Gibbons, from wood, in imitation of point lace, and was so finely worked, that it could be folded and tied. Gibbons gave it to the Duke of Devonshire on completing his labours at Chatsworth.

CHAPTER V.

Gloves on the Bench.

WHITE gloves at a maiden assize represent the zero of crime—the antithesis of the black cap. They afford a distinct foretaste of the Millennium; the occasion of their presentation is held to reflect credit on any town or neighbourhood, and is widely noticed in the newspapers. The judge says pleasant things to the jury, the foreman of the jury thanks the judge and says more pleasant things, and everybody is delighted. There are some places in which these little episodes are not rare. Recently, for instance, the Recorder of Cambridge was, for the third time in succession, presented with a pair of white gloves because "nobody had done nothing." It has been slanderously stated that when business, from a police point of view, is not brisk, there is no scruple in holding over a case or two for a following sessions or assize, but this is hardly to be credited.

Of this ancient institution—the chief remaining relic of the historic past of gloves—the origin cannot be decidedly traced. It is, of course, most feasible that the colour of the glove tendered is intended to perpetuate the memory of an unusual innocence, or absence of crime, but the practice is carried into other than legal circles. A few years ago, for instance, no sales having on a certain day been effected on the Liverpool Stock Exchange, the president of the Stockbrokers' Association was, with some formality and many complimentary speeches, presented with a pair of white gloves. The occasion is more than usually memorable on account of a description of the incident being headed in the next day's *Times*, " Diminution of Crime in Liverpool."

The custom, which has similar currency over the Border, appears to have acquired a far wider latitude than was, at a former period, allowed to it, for the white gloves were only presented to a judge when he presided over an assize at which no prisoner was capitally convicted, and this alone was known as a maiden assize. This disposes of a plausible suggestion that the gift of a pair of gloves to a judge was tantamount to saying that he need not come to the Bench, because it was once prohibited to judges to wear gloves when presiding at an assize—his services were not required, he might go and wear gloves. Maiden assizes of the ancient order were, under the blood-thirsty penal code formerly in force, far less common than the maiden assize as we know it now. It is startling to remember that there used to be no less than 223 distinct offences punishable with death, and more so to remember that reform, on this matter, has been brought about within the reign of her Majesty, our Queen. In the first year after her accession, the category of capital offences was brought down to seven. Three years before that, in 1834, there were 480 executions, of which only about one-fifth were for offences against life or limb. The maiden assize has altered its character with the change in the law. Brand, in his *Popular Antiquities*, explains it, " when no prisoner is capitally convicted." The custom is traced farther back by a passage in Fuller's *Mixt Contemplations on these Times*, 1660, which says, "It passeth for a generall report of what was customary in former times, that the Sheriff of the County used to present the judge with a pair of white gloves, at those which we call Mayden Assizes, *viz.*, when no malefactor is put to death therein." Another book of the same century makes it appear that a similar present was offered by such prisoners as received pardon after condemnation. Clavell, a highwayman who was granted the king's pardon, wrote, in penitence, a *Recantation of an Ill-led Life*, published in 1634, and dated from the King's Bench Prison, October, 1627. The dedication is addressed "to the impartiall Judges of his Majestie's Bench, my Lord Chief Justice, and his other three honourable assistants," and contains these lines :—

> Those pardon'd men who taste their Prince's loves,
> (As married to new life) do give you gloves.

The reversal of the process of outlawry, which might, in some measure, be considered as a living death, was also followed by a present of gloves to the sitting judges who restored the man to all the pleasures of home, and gave him again the privileges of a citizen and the protection of his king. The outlaw was compelled to appear in person to beg for his restoration to society as fitting punishment for having refused to answer the five several calls which had "proclaimed, required, or exacted" his appearance after the sheriff had, in the first instance, failed to find him. An old *Law Dictionary* of 1670, says, "At present, in the King's Bench, the outlawry cannot be reversed unless the defendant appear in person, and by a present of gloves to the judges implore and obtain their favour to reverse it" (BLOUNT) ; and this custom had acceptance long before ; for in an instance given in the *Year Book*, 4 Edward IV., the culprit "paid the fees of gloves to the Court, two dozen for the officers of the Court (for these in all four shillings), and in addition three pairs of furred gloves for the three judges there, to wit, Markham, Chief Justice, Yelverton, and Bingham." By this time, at any rate, the judges had acquired the privilege of wearing gloves. Hollar's engraving of the coronation procession in 1660 shows one of the judges wearing a pair of gloves richly fringed about the cuffs, and the portraits of the judges painted by order of the Corporation in the reign of Charles II., and hanging in the Courts at Guildhall, represent them with fringed and embroidered gloves. In the preceding century, too, Sir Thomas More, according to a well-known anecdote, refused the lining of a pair of gloves offered to him by a grateful suitor who had won her cause before him. The "lining" consisted of forty angels, and, with the glove in which they were placed, was offered on the New Year's Day following the decision, as the custom was. "Mistress," wrote the upright judge in reply, "since it were against good manners to refuse your New Year's gift, I am content to take your gloves, but as for the lining I utterly refuse it." We may take it for granted that it was not every judge who had the probity and integrity of good Sir Thomas, and there may have been no remote relationship between this incident and the old injunction contained in the *Speculum Saxonicum* (lib. iii.), that judges were not to wear gloves on the bench. It is significant that any bribe came to be known as a pair of gloves ; to have given a pair of gloves did not always actually denote a present of gloves, but some equally valuable gift tendered in return for services rendered. Grose quotes from an old play an incident where an alderman presents his glove filled with gold pieces as a bribe to procure a commission for his son. The Portuguese have a proverb on this head, expressive of a person's integrity, *Não traz lavas*—He does not wear gloves.

According to Mrs. Bury Palliser (*History of Lace*), judges at maiden assizes were presented with "laced gloves," instancing the gift by the sheriff of Lincoln at the Lent Assizes to Lord Campbell, the presiding judge, of a pair of gloves "richly trimmed with Brussels lace and embroidered with the city arms embossed in frosted silver on the back."

The custom is, in its incidence, variously followed. The judge always receives a pair, but in some instances, all the officers of the court are given either gloves or an equivalent in glove money. The sheriff is invariably the giver of the gloves. It is merely a supposition, and the suggestion is offered with some diffidence, but may not this presentation of gloves by the sheriff at a maiden assize—taking into consideration the original acceptation of the term, an assize at which no prisoner was capitally convicted—have some connection with the horrible office of executioner, which has been the duty of the sheriff when no person willing to take the office for the sake of reward or the remission of sentence could be found ?

Gloves had further connection with legal observances in an ancient custom requiring that, when the judge invited the justices to dine with him at a county assize, a glove was handed about by the clerk or crier of the court, into which every person invited put a shilling (PEGGE) ; and still have, in the necessity that persons taking the oath should first remove their gloves. This obligation was dispensed with by Baron Bramwell, at the Liverpool Assizes in 1838, when a witness before him, having some difficulty in removing his glove, was allowed to be sworn, " holding the sacred volume in a gloved hand ; " the learned judge remarking that, " he knew no reason why a witness should be required to remove his glove when taking an oath " (TIMBS). There has, however, always been a recognized difference between the gloved and ungloved hand. It is held disrespectful to come into the presence of the sovereign with gloved hands, as it was once obligatory to doff gloves on entering a church. It was a common mark of respect ages ago to doff, or do of, the gloves. *The Boke of Curtasye*, one of the earliest of manuals of etiquette, ascribed to the middle of the fifteenth century, directs him who would be truly courteous, to " do of thy hode (hood), thy gloves also " on entering the hall door of a house. Professor Thorold Rogers says that the proffer of a bare hand was in ancient days a symbol of hostility, that of the gloved hand a token of peace and friendliness, and, although the conditions are reversed, the fast-fading custom of taking off the glove on shaking hands with a friend doubtless had its origin in this ready indication of amity or enmity. The practice of presenting gloves to visitors by the Universities, is said, moreover, to have been an indication on their part that they recognized in their guests such dignity or learning as made them worthy of remaining with covered hands even in the presence of the highest collegiate dignitaries.

There were thus several reasons prescribed by the *lex non scripta* of ancient usuage—even though they might not be found in Blackstone or Coke—for differing from the conclusion of the learned judge, whose precedent has happily been disregarded ; for it certainly would not be wise to lose any of the hold the oath has upon those who take it, nor to drop any of the formalities which give additional weight to a solemn ceremony which is, as can be gathered from our newspapers or, better, by an occasional attendance at our courts of justice, too often lightly esteemed and set at nought of the people.

CHAPTER VI.

Ibawking Gloves.

IN hawking, our forefathers reduced a sport to a science. It had its own peculiar vocabulary in the which its votaries were proud to be proficient, and to it were devoted exhaustive treatises in the infancy of our literature. Hawks were the constant companions of people of rank, an essential part of a great man's retinue, and that the most important. No man, however high, thought it scorn to feed and tend the birds which were deemed fit presents from one king to another. Commercially, they were considered beyond price, and only parted with at enormous sums as a particular favour:—

> For empty fistes men used to say,
> Cannot the Hawke retaine.
>
> *Boke of Nurture*, 1577.

They were protected by most severe laws. It was penal to steal or conceal a hawk; and once, in the time of Edward III., the Bishop of Ely excommunicated certain persons who stole a hawk from her perch in the cloisters of Bermondsey. A double palliation of the sentence is offered: the theft was committed during Divine service in the choir, and the hawk was the property of the bishop. Considering the fervour with which ecclesiastics followed falconry, and

" loved venerie," as did Chaucer's Monk, setting at nought all prohibitions and regulations to the contrary, it is to be feared that the latter reason far outweighed the former.

It would be foreign to the purpose of this book to trace the origin of hawking from its birth in the fourth century to its decline, or to offer any lengthened dissertation upon its practice. It is sufficient to remark that the birds were at one time considered emblems of dignity, and their possession confined to landed gentry and noblemen. It was held dishonourable to part with this distinguishing mark of rank. Captives endured their prison rather than purchase release by relinquishing their hawks ; and the laws of Charlemagne expressly forbade that hawks should form part of a ransom. So inseparable were the birds from their owners, that men chose to be represented with hawk on fist, and "the ancient English illuminators have uniformly distinguished the portrait of King Stephen, by giving him a hawk upon his hand," to signify, Strutt thinks, that he was nobly though not royally born. An Irish king of the same century is, in some rude frescoes in the Abbey of Knockmahoy, county Galway, represented crowned and with a hawk on his gloved hand.

It is a fair inference to suppose that the wearing of gloves has been coexistent with the pursuit of hawking ; cleanliness, no less than some protection from the sharp talons of powerful birds, must have demanded some hand-covering when carrying them. The birds were always borne on·the hand, so that, when unhooded and the jesses loosened, the prey might be the more readily sighted, or that they might be petted and caressed. And, although it is remarked that Harold is depicted on the Bayeux tapestry carrying a hawk on his ungloved hand, this may but arise from the one great difficulty which hinders the study of gloves—the fact that gloves, unless of a peculiar shape, or represented of a particular colour in illuminations, are, from their fitting the hand tightly, not to be easily distinguished from the hand itself. It is worth notice that in one of the earliest illustrations that we have of the employment of anything like a glove, the wearer is represented as carrying some birds in her hand (*ante, pp.* 13, 14).

In the thirteenth century ladies adopted the sport of hawking, and became soon so proficient as to be considered not only as equal with, but actually superior to their lords. Strutt, in his *Sports and Pastimes*, engraves from a MS. written early in the fourteenth century, and preserved among the Royal Collection, a group of ladies following this favourite pastime, one of whom carries in her hand a hawking glove, and another bears a hawk on her gloved fist. This encroachment of females on a sport hitherto the monopoly of males, in an age when woman was but held by her husband—

Something better than his dog, a little dearer than his horse,

and probably not nearly equal in value to a ger-falcon or a tercel, appears to have been a matter of much wonderment. These masculine ladies were freely satirized and sternly rebuked by contemporary writers, who, at the same time, inconsistently condemned the pastime as wanting in robustness and too effeminate for men to follow. Still hawking did not decline, but rather increased in popularity. Old Robert of Gloucester—a monk who wrote a rhymed History of England some time in the thirteenth century— says of a gentleman, " neuer but whenne he bereth hawkes ne veseth he gloues." Kings continued to follow the chase,

HAWKING GLOVES.
(In the possession of Col. J. S. North, M.P., D.C.L.)

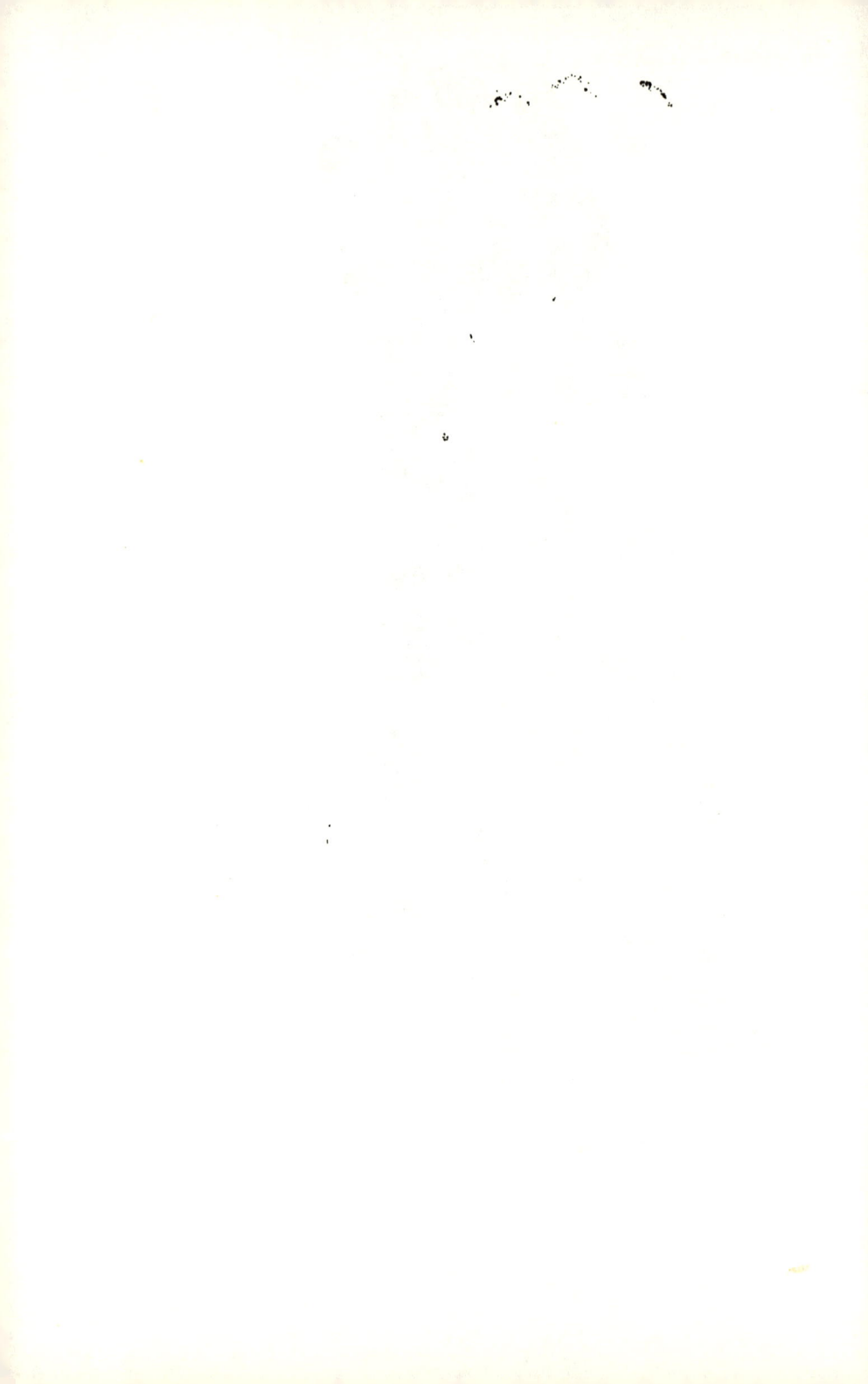

and Richard II. kept his hawks at the famous Mews at Charing Cross, when Mews meant, not stables, but in falconers' phrase, a place where hawks were kept at moulting time. "The Mues," in 1537, were made into "the beautiful stable for the King's horses," of which Camden speaks, and so preserved in a perverted sense a word which would otherwise have become obsolete.

This conversion did not denote any neglect by the Eighth Henry of the sport which his ancestors had always loved well. Henry, although he is, in his commonly received character of a regal Blue Beard, only one of many historic ninepins, was certainly a manly king, fond of tennis and athletic exercises, and certainly fond of hawking. In his *Privy Purse Expenses*, there is in

HAWKING GLOVE OF HENRY VIII.
(ASHMOLEAN MUSEUM).

October, 1530, a payment of Twenty Shillings, "paied to a servaunt of Maister Bryan in rewarde for bringing of a crosbowe, a quyver w^t arrowes, and a hawkes glove." Another payment is for "iv hawk gloves a^t vj^s viij^d le glove." In an inventory of his effects at Greenwich taken after his decease there appear, "Two quivers with arrowes," "Forty-four dog collars of sondrye makynge," "Twenty nine bowes," "One hundred and thirty eight hawkes hoodes," "Three payre of hawkes gloves, with two lined with velvett." At Hampton Court, in the Jewel House, were "seven hawkes gloves embroidered." (*Harl. MS.*, 1419.)

There is in this inventory an item which foreshadows the decline and final extinction of falconry. "A gonne upon a stocke wheeled." Fire-arms, although they had been introduced a century before, gave a death stroke to hawking and archery when they came into general use, and both pastimes quickly decayed.

Gloves were likewise used in archery, another sport in which ladies made themselves proficient. Gage's *Antiquities of Hengrave* show a payment, in 1574, of sixpence, for "ij shooting gloves for my mistres;" and again in the *Privy Purse Expenses of Henry VIII.* is an item: "The same daye paid to Scawseby for bowys, arowys, shaft (brode hedd), braser, and shoting gloves for my Lady Anne, xxiijs· iiijd." In 1564 Richard Seymour in his will bequeaths to "Thos· Shaw one handkercher, a washing ball, my chest (chess) bord and chest men, and my red gloves;" and Robert Midforthe leaves "one bow, a shef of arrowes, and a shoting glove" (*Wills and Inventories: Surtees Soc.*) So, in the *Shuttleworth Accounts* (*Chetham Soc.*), tenpence is paid, in 1612, for "two shooting gloves for archery."

By the beginning of the seventeenth century hawking had experienced that surest sign of decay in fashion of any kind—it had become common. Bishop Earle, in his *Microcosmography*, 1628, scoffs at the upstart country gentlemen for esteeming a hawk "the true burden of nobility," and for being "exceeding ambitious to seem

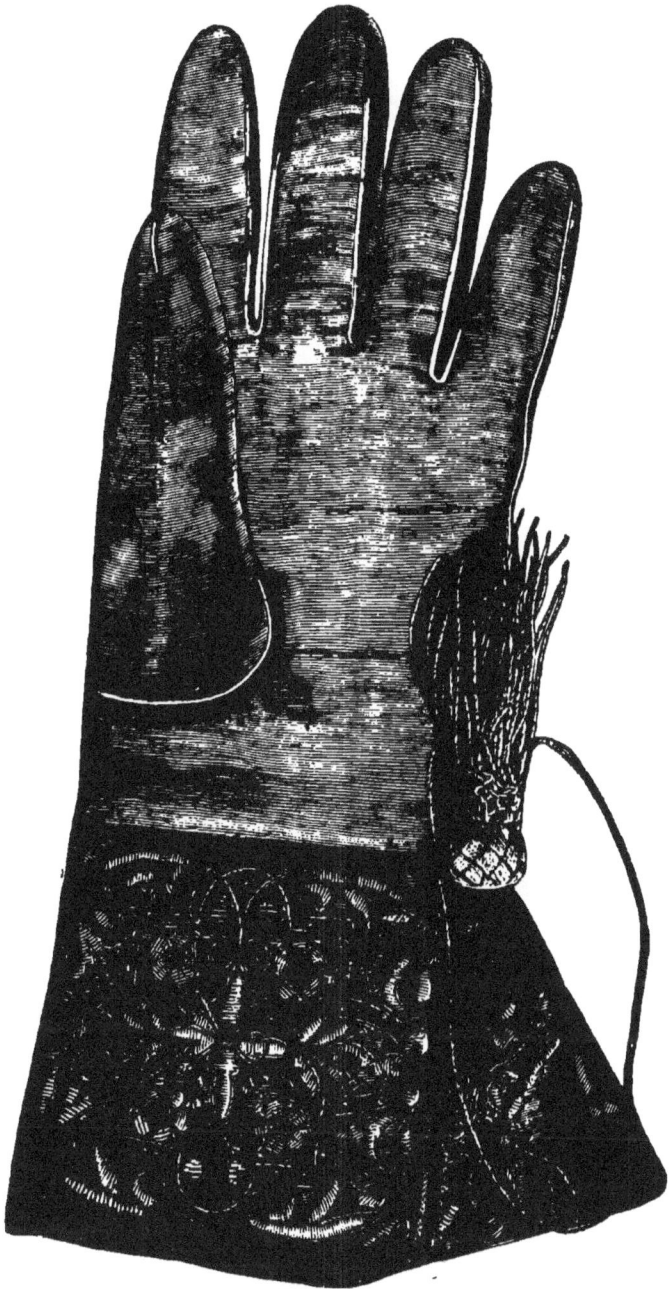

HAWKING GLOVES.
(In the possession of Col. J. S. North, M.P., D.C.L.)

F

delighted in the sport and have his fist gloved with his jesses." By the end of this century, says Strutt, the sport was rarely practised, and a few years afterwards hardly known. In 1662, when the Russian ambassador brought Charles a present of rich furs, carpets, cloths of tissue, seahorse teeth and hawks, he took two or three of the hawks upon his fist, but donned a glove wrought with gold, which, Pepys says, was given him for the purpose. It was no part, then, of a king's wardrobe.

There are preserved at Wroxton Abbey, Oxon, two splendid specimens of hawking gloves, said to have been worn by James I. when he visited the Abbey to stand god-father to one of the children of the North family, and was entertained with sports, hawking, and bearbaiting. If he wore these gloves he was right royally equipped. There are *en suite* a hawking pouch and lure (of which fine engravings are given in the *Archæological Journal*), and the whole were exhibited by Lady North in 1862 before the Archæological Institute. The design of the embroidered pouch corresponds with that of the gloves, and consists of a trail-branching pattern formed of the blackberry in flower and fruit, and the mistletoe ; possibly symbolical of the season at which the disport of hawking was most in vogue.

Chapter VII.

Gauntlets.

IT argues no small courage, not to say hardihood, for a writer in these days to take up a question of armour. Much of the old spirit of chivalry remains yet in those who make a study of the trappings of ancient knights. It may be that these literary champions become imbued with the touchy dignity, pride, and unquestionable courage which led doughty warriors to seek danger wherever it might be found, in a chance encounter, in a crusade, in foreign service, or even in a tournament. *They* scented the battle from afar off, and so do those who love them as a doctor does a "subject." Does anybody hunger after a new pleasure? Let him study for a few years the question of the introduction of chain-mail, or become learned in helmets, with the minor puzzles of visors, basinets, aventailles, volantes, armets, unibers, bevors, and other accessories, and then make known the theories he will assuredly originate. Life will soon have a fresh zest as keen antagonists crowd around, each fighting for his own hand, but all ready to break a lance with the bold new-comer; and for the remainder of his term of years there will be no lack of excitement, no fading of interest for him. Alexander himself would not have wept had he lived in these days and become an armorial antiquary.

Happily there is very little dissension about gauntlets ; their descent is tolerably clear. It will of course be understood that gauntlets have not here the modern interpretation given to them, when, by some strange etymological evolution, the term denotes, instead of "war gloves," as they are called in Scotch statutes of the fourteenth century, the wristpiece or cuff of a glove. Many dictionaries, indeed, define them as iron gloves, Dr. Johnson's included— which is nearer the truth, but only in part. They were in the thirteenth century written down "*chirothecis ferreis.*" Originally "gauntlet" and "glove" were synonymous, and under its mediæval form of *wantos* gauntlet often does duty for its more peaceful relative. *Wantos* is not the only form given to the word. Ancient scribes were each, as regards orthography, a law unto themselves ; and in this instance sometimes wrote *guantus, wantus, wantonus,* or *wantonene.* There is little doubt that the word comes to us through the French. Bede speaks of it as a French term ; and another use of the word in the seventh century occurs in the life of St. Columbanus written by Jonas, Abbot of Bobbio, in which he writes, *Tegumenta manuum quæ Galli wantos vocant.* Isaac Taylor, in his work on *Words and Places,* would have had us derive the name and the manufacture together from Gaunt or Ghent in Flanders. This holds good in numberless instances, and is of undoubted value in determining the origin or tracing the progress of manufactures. In reviews of the work the author was twitted with an antiquity of the word which was supposed to at once convict him of an error ; not only was it used in legal formulæ of the ninth and tenth centuries, but Bede spoke of it in the seventh, and this was held to be conclusive condemnation of the association between Ghent and gauntlets. It might have been retorted that Ghent was at any rate of equal antiquity, and was also spoken of as an established city in the seventh century. Still there is no evidence of the manufacture of gloves ever having been particular to Ghent; and the derivation of the word through the French *gant*, the hand, is too obvious to be disputed. The word, too, is also found in the Icelandic *vöttr* and the Danish *vanter*, certainly carrying it beyond Mr. Taylor's ingenious theory, which was allowed no part in the second edition of his work.

The gauntlet, as a distinct article of defensive dress, did not come into use until the thirteenth century, but they were then only the outcome of progressive improvements. Olaus Magnus, in his *History of the Northern Nations*, speaking of armour, says, "Anciently they wore heavy helmets rudely fashioned according to the art of that age, and thick tunics made either of iron, leather, or felt, lined with linen or wool; also iron pieces for the arms and gloves." Indeed, it would seem to have been most improbable that in those days of close fighting so vital a member as the hand could have gone unprotected. But among the Anglo-Saxon warriors, so far as illuminations guide us, the hands were not covered. They appear to have trusted more to their own prowess than they feared that of their enemies, paying little attention to defensive armour, and, with the exception of a shield and helmet, fighting in their ordinary dress. It may have been, then, that the use of gloves would have been considered a despicable weakness. Later, after the Norman invasion, the sleeves of the hauberk, which, on the Bayeux tapestry, stop at the wrist, were lengthened to cover the hand, an oval-shaped opening on the palm allowing of the hand being withdrawn at pleasure. These hand-coverings had no divisions for the fingers, an improvement not introduced until the reign of Edward I. To these sleeve "continuations" succeeded plates, entirely covering the back of the hand, and made flexible opposite the knuckles. These plates had loops through which the fingers were thrust. At the same time. gauntlets—mail-covered gloves—first made of overlapping plates of metal, began to be worn, consequent upon sleeves ceasing at the wrist. These gauntlets, according to Planchè, had a leather foundation, the exterior coated with scales or other formed pieces of plate. "Some of leather only are seen on the effigy of Dubois, engraved by Stothard, and in the mutilated effigy in St. Peter's, Sandwich." Waller, in his work on Brasses, engraves a scaled gauntlet from the brass of Sir Richard de Burlingthorpe, and offers the suggestion that the scales may have been of horn or whalebone. Gloves with gauntlets of scale work were worn so late as the time of Charles I. Planchè also copies from a drawing of M. Viollet-le-Duc, three views of a leather gauntlet of the thirteenth century, in which the vambrace, or fore-arm armour, is brought up over the knuckles to give additional security. Some gauntlets of this period were made wholly of steel, and the wearer first covered the hand with some soft material to prevent the metal abrading or bruising the flesh, even as they wore counterpointed garments, stitched *pointe-contre pointe*, across and across in panes over wadding, like an old-fashioned counterpane, to keep the armour from being an unfriendly protection. A French poem of 1296 says that knights went with their hands—

Covered by stuffings of wool
And gloves of plate riveted,
Pierced with holes in many places.

"Gloves of plate" is the term used for gauntlets in a complaint made, in 1328, by John Bully of Nottingham, against John de Melton of Nottingham, that from him was detained unjustly a horse of the value of 15s., and "a haketon, a haubergeon, and a pair of gloves of plate, a bascinet, of the value of 20s." (*Records of the Borough of Nottingham*). In this plaint "the said John" did not prosecute, or, if the case had been further heard, we might have been able to know the separate value of the gloves of plate for which John de Melton had so unlawful a love. In the reign of Edward III. arose a new fashion, of fastening projecting spikes of steel to the several knuckles, giving additional emphasis to the assertion that the wearer was armed at all points. These were styled "gadlyngs," not inaptly, the name coming from the old goad which our Anglo-Saxon forefathers called *gaad*, and used to drive their cattle with. Their introduction was dramatic enough. In a quarrel between Hugh the Valiant, King of Cyprus, and the King of France, either party appointed champions to defend their cause in an appeal to arms, the Cyprian King nominating Sir John de Visconti as his champion, and Sir Thomas de la Marche appearing for the King of France. The combat was appointed to take place before the English King, and on a set day the two delegated antagonists appeared before him in the lists, with the Prince of Wales and all the Court looking on. At the sound of trumpets the combat began. At the tilt both their spears broke on the opposing shields without dismounting either, on which they alighted to continue the fight with swords on foot. Each found in his antagonist a foeman worthy of his steel, for after fighting a considerable time without either gaining any advantage both their weapons were rendered useless. They came to close quarters, grappled with each other, and fell locked together. The visors of their helmets were defended with small bars of steel, through which they might see and breathe more freely, all the rest of their bodies being completely covered with armour. " Rising together," says the account of this notable duel, " Sir Thomas got the advantage of his antagonist by having sharp pricks of steel, called gadlings, fastened between the joints of his right gauntlet, and therewith struck at the visor of Sir John who had no gadlings on his gauntlets. Striking as often as opportunity offered, he grievously hurt him in the face, so he exclaimed that he was unable to help himself. At this King Edward threw down his wardour, the marshal cried " Ho !" and the combat ceased."

The victory was, of course, with the French knight, to whom the vanquished was handed over to be dealt with at discretion. Sir Thomas did not exercise his power of punishment, but chivalrously presented his prisoner to the Prince of Wales, who at once gave him his liberty, and so the matter ended. (MEYRICK : *Antient Arms and Armour.*)

This event caused gadlyngs to be commonly worn, and the introduction of these knuckle spikes added greatly to the ornamental character of gauntlets—each point being surrounded by rich ornaments, which were continued on other portions of the glove, until the whole, from wrist to finger tips, was overloaded with fine raised or engraved designs. On the gloves hanging still over the tomb of Edward the Black Prince, in Canterbury Cathedral, may be seen some curious gadlyngs, like small erect lions, formed of brass, the ordinary spikes only appearing on the first joints of the fingers. The lions are very probably only fanciful or heraldic embellish-ments made on gloves never worn by the Prince, but solely for sepulchral service. Such armour was often made, and many of the gauntlets which, with shields, swords, and spurs — sometimes,

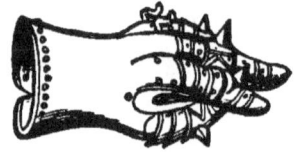

GAUNTLET OF EDWARD
THE BLACK PRINCE.

even boots—are so frequently found over the resting-places of departed knights, are, so far as their personal associa-tions are concerned, very often venerated without cause. In many instances they are of too small a size to have ever seen active service. The gauntlets on the effigy of the Prince, which are far more likely to be faithful representations of his actual armour, have no miniature lions upon them.

Some instances have been found of fifteenth century gauntlets being enclosed by steel mittens. These have invariably articulated fingers, a somewhat unusual feature in gauntlets at that period, the back of the hand being again covered by overlapping plates of steel, sometimes plain, but frequently fluted — still oftener, splendidly engraved. Hinges and clasps were employed to enclose the thumb, and fasten across the front of the hand. They were then commonly called gloves, although either term was employed, for gauntlets are directed in

GAUNTLET—1450-70.

the reign of Henry VII. "to be provided for the King's henxmen, and to the master of the same" (*Record Papers: Materials for a Hist. of the Reign of Hy. VII.*) In Harrington's *Nugæ Antiquæ* is quoted, from a volume once belonging to Sir John Paston, and now in the Lansdown MSS., a curious list of articles of armorial dress, in the order of putting them on. The preliminary processes consisted of enveloping the warrior in a doublet of fustian, "kut full of holis," which is enjoined to be "streightly bounde;" next, after adding "gussets of mail," dressing

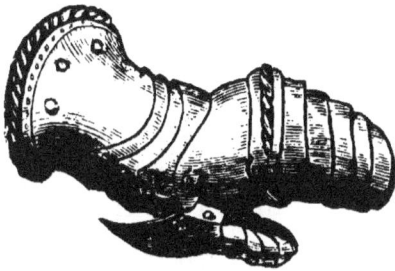

GAUNTLET, 1525.   GAUNTLET, 1535.

him in "a pair of hosen of stamyn single" (stamen being a kind of serge), adding "a pair of short Bulworks" and "a pair of shone." *Thus far in preparation*, says the writer, and indeed the work of arming a man must have occupied a considerable time, for no less than fifteen distinct stages are shown after the preparation had concluded. First, come the "sabatynes," or steel clogs; next, the "griffus," or greaves; the "quysshews," cushions for the knee caps; then the "breche of maile," armour about the loins; then "towletts," little tiles or pieces of steel overlapping after the manner of tiles; then, in succession, the breste (breast-plate), vambrace (armour covering the wrist to the elbow), and rere-brace (that covering the elbow to the shoulder), and "then the cloovis" (gloves). After this the dagger, short sword, and long sword had to be hung in place, the "cote" and "bassenet" (helmet) had to be severally added

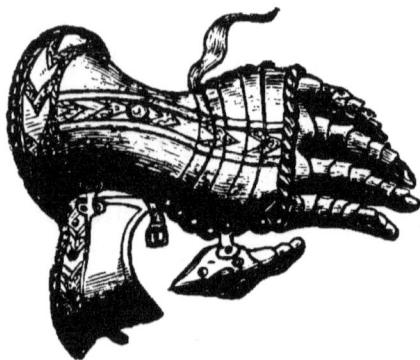

GAUNTLET, 1543.

before the "Pensell" or little pennon could be put into the hand of the completely-armed knight. We may well wonder how, so stuffed, braced, and encumbered, he could manage to move or breathe under the weight and restriction of his defensive equipment, to say nothing of fighting.

In the *Calendar of State Papers* (Scotland) is shown a letter from John, Lord Darcy, in 1572, to Lord Burghley, informing him of the detention of a Scotch ship, sent from London, which contained "two double bases and two single bases of iron, without chambers, also corslets, callivers, dagges (pistols), arming gloves, and three small barrels of powder."

Gauntlets continued in use during the Stuart period, when they were, as at first, worn with continuations to the elbow, some being shaped to the arm, but others being merely cylinder coverings. The fashion is said to have been then borrowed from the Asiatics. Some were made of overlapping pieces of serrated leather, covering the back of the hand like a shield ; some of layers of leather and loose pasteboard ; and others of the quilted silk—in which the English Solomon felt more secure against the assassination he so greatly feared, and which was worn so commonly by people who, in his time, went about in dread of a Protestant massacre, making themselves look, in silken " backs," and " breasts," and head-pieces, like so many hogs

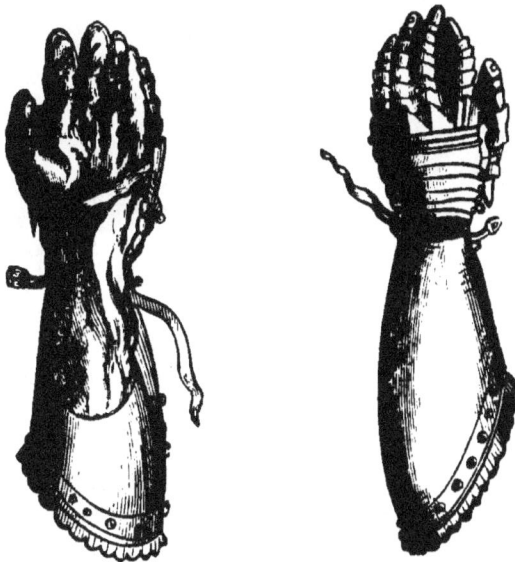

INSIDE AND OUTSIDE OF LONG (OR ELBOW) GAUNTLET.

in armour, says Roger North. In 1631, commissioners appointed by Charles I. to settle the prices for making and repairing armour, and to secure uniformity in the fashion of armour and arms, fixed the price of making "a gauntlet glove" at 3s. 6d. ; but soon after this gauntlets proper fell into disuse—the "buff-glove," which appears to have originated with Cromwell's troopers, affording sufficient protection to the hand and wrist. The heavy cavalry still wear such gauntlets, although of a lighter character than they were at first made, when to hands of sheepskin were added long tops of stout buffalo hide, coming half way up the arm, and sometimes tops of scale work.

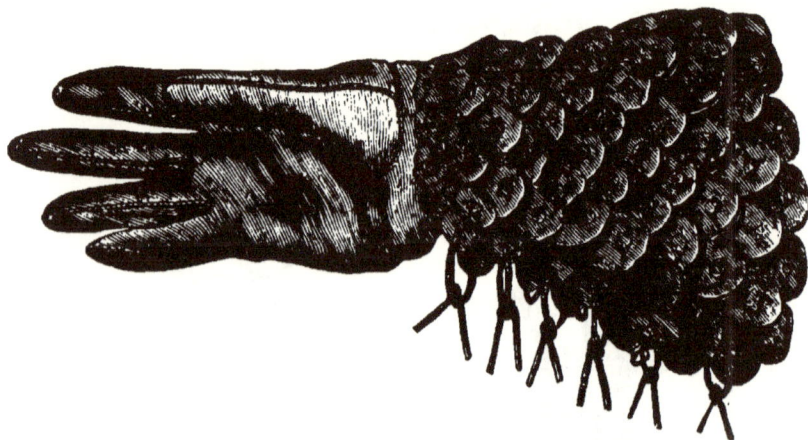

BUFF GLOVE OF SCALE WORK (TOWER ARMOURY).
(From the *Archæological Journal.*)

The ceremony of conveying a challenge by casting a gauntlet seems to have been observed in the less deadly, but still dangerous, justs and tournaments which were such attractive shows in the middle ages. This, considering the elaborate ceremonial with which such occasions were inaugurated, and the very close resemblance in their conflicts to actual warfare, is no more than was to be anticipated; but, considering the significance attached to the action of defiance by casting the glove, it goes to prove yet further how much of feeling and personal antagonism was imported into these very rough sports. Before fighting, the combatants took oath that they had used no undue advantage, of witchcraft or subtlety, or any unfair device to procure victory. One of the subtleties thus guarded against was the close gauntlet, of which Hall makes mention—that is, gauntlets with immovable fingers, with the exception of one joint, so that in the shock of meeting an adversary, the weapon of the wearer, held fast in the fixed fingers, might not be dropped out of the hand.

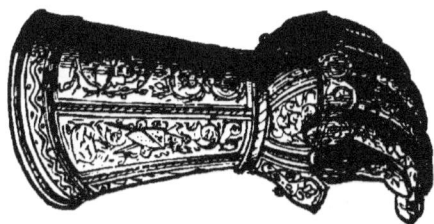

CLOSE GAUNTLET, TEMP. HENRY VIII.

Although the lists were only open to knights and esquires, the mimic warfare had such hold upon the affections of the people, and was so consonant with their chief pride, that apprentices and citizens held an annual tournament of their own. In Fitzstephen's time this was fixed for each Sunday in Lent; and young noblemen, not yet knighted, did not disdain to come and make trial of their skill with the commoners.

At all times the quintain afforded an outlet for the martial aspirations of the people. This was sometimes a fixed mark to run a-tilt at, more often some object on a pivot, which, not being fairly struck, would give the faulty marksman a heavy blow as he rode past. But this exercise, even though it required no small dexterity, was improved upon by running spear in hand at a fully-armed man, who, with a shield, parried as best he could the strokes directed at him, but acted only on the defensive. Du Cange quotes from an ancient author (*Le Roman de Giron le Courtois*), who introduces one knight saying to another, " I do not by any means esteem you sufficiently valiant for me to take a lance and just with you ; therefore I desire you to retire some distance from me, and then run at me with all your force, and I will be your quintain." In the *Byting Satyres* of Bishop Hall (Book iv., Satire iv.), Gallio, typical of pretentious vapouring coxcombs, is thus advised :—

> Pawne thou no glove for challenge of the deed,
> Nor make thy Quintaine others' armed head
> T' enrich the wasting herald with thy shame,
> And make thy losse the scornful scaffold's game.

It would seem from this that it was customary, in running at a living quintain, to offer a challenge even in play by means of the glove, and that the herald who, in close imitation of nobler sports, proclaimed the joust and watched its course, was entitled to the pledged glove in case of failure. If this be so, then the glove, as the emblem of man's honour and evidence of his courage, had a common and universal acceptance among all ranks of the people.

In these sports we find that the glove was not only a forfeit from the unskilful, but a reward of success. The *Calendar of State Papers* (Venetian) from a letter in St. Mark's Library, gives particulars of a stately joust performed "on the day of the Magdalen," in 1523, in front of the apartments occupied by the Emperor and the Queen of Portugal. "The jousters were twenty in number— ten on each side. On one side was the Emperor, with nine other lords and gentlemen, all unmarried ; their opponents being the Duke of Najera, and as many more lords and gentlemen, all married. The liveries were of two sorts, and very magnificent. The appointed prize was forty pairs of *perfumed gloves*, worth upwards of two hundred ducats, but it has not yet been awarded, as the jousters did not finish their courses. The first spear was run by the Emperor, who in truth becomes daily more and more admirable in military exercises."

Other athletic exercises of lower degree, were rewarded wholly or in part by gloves. An instance is given in *A Mery Geste of Robyn Hode* (included in Garrick's *Collect. Old Plays*), where a bull, a courser with saddle and bridle, a mug and a pipe of wine, and "a payr of gloves," were all offered as prizes to the best wrestler that presented himself.

CHAPTER VIII.

Perfumed Gloves.

ONE of the most generally received statements as to the history of gloves is that the first perfumed gloves brought into the country were those given by Edward Vere, Earl of Oxford, to Elizabeth, in the fifteenth year of her reign, after he came back from Italy. This account, given on the authority of Stow—honest John, the Chronicler—is met with again and again in studying gloves, sometimes—as in Walpole's *Royal and Noble Authors*, for instance—varied by showing the gloves embroidered instead of perfumed, but still making the assertion positive that at this date they were first introduced into the country. What Stow really did write was this :—

" Milloners or haberdashers had not then any *gloves imbroydered*, or trimmed with gold, or silke ; neither gold nor imbroydered girdles and hangers, neither could they *make any costly wash* or *perfume*, until about the fifteenth yeere of the queene, the Right Honourable Edward de Vere, Earl of Oxford, came from *Italy*, and brought with him gloves, sweete bagges, a perfumed leather jerkin, and other *pleasant things ;* and that yeere the queene had a *pair of perfumed gloves* trimmed only with four tuffes, or *roses of coloured silk*. The queene took such pleasure in those gloves that she was pictured with those gloves upon her handes, and for many years after it was called ' *The Earl of Oxford's perfume.*' "

There is, in this, no warrant or authority for the inaccurate conclusion based so commonly upon it; it shows only that the pompous Earl brought back, after his seven years voluntary exile, sundry personal refinements, among which, as we take it, there were a pair of perfumed gloves. These were not the first of the kind known here by any means, as Stow would, in all likelihood, have been well aware. Ample evidence of perfumed gloves being known here before Elizabeth's time can be adduced. In the thirty-second year of the reign of Henry VIII., as appears from a *Book of Quarterly Payments* for his household, Arcangell Arcan, Gunner, did, on New Year's Day, present his Majesty with a pair of perfumed gloves, receiving twenty pence in reward. In 1532, according to the same king's *Privy Purse Expenses*, there was "Paied to Jacson, the hardwareman, for a dousin and a halfe of Spanysshe gloves," 7s. 6d. ; the gloves of Spain being famous even then for the scent skilfully imparted to them. So also, in 1544, there was "given to Master Wheller's servaunt for bringing " a pair of swet (sweet = perfumed) gloves" the sum of 2s. (*Privy Purse Expenses of the Princess Mary*), and in the same year a gratuity of thirty shillings was given to "a gentleman of my Lord Admyralles bringing a coffre w^t. x paer of Spaneyshe gloues from a Duches in Spayne." This coffer was in all probability a box of aromatic wood, in which such articles were kept. The "Mistress of the Sweet Coffers" had a recognized place in royal establishments, probably subordinate to the Mistress of the Robes, and had in charge minor articles of dress in caskets of cedar or other fragrant woods. Queen Charlotte, says Pegge, kept her gloves in a perfumed box.

The "sweete bagges" introduced by Vere were later

brought to great splendour. For instance, in an Inventory, taken in 1614, of the effects of Henry Howard, K.G., Earl of Northampton (*Archæologia*, xlii.), there appear several most elaborate sweet bags, splendidly embroidered, some with " flies, wormes, and flowers," and others necessitating full description.

> Item, two verie large sweete bagges, embroidered with embosted worke of silver, gold, and coùloured silkes, and filled upp with ovals of divers personages, lined both with clowded sattenes, black, the ground white £16

> Item, a smaller sweet bagge embrodred with high embosted mosseworke havinge two sea nymphes upon dolphins, and other figures of fowles, edged about with lace of silver and gold lined with carnation ... ... 50s.

For further evidence as to the introduction of perfumed gloves there were again, in 1556, among the gifts, or " guiftes," as the precarious orthography of the period preferred it, brought to the Queen's Majesty, on New Year's Day, from " Mr. Frankewell, a paire of perfumed gloves," from " Pascal, a peire of gloves perfumed and cuffed with gold and silver," and from " Hannyball, a peire of perfume gloves." These latter items are to be found in *Expences of Antient Times*, where, with a full and lengthy list of gifts, they are quoted from " the original Roll, formerly belonging to Sir Wm. Herrick, of Beau manor, and still in the possession of his descendant, William Herrick, Esquire." These facts are quite conclusive as to the introduction of perfumed gloves prior to the days of Good Queen Bess ; and, in addition, it would have been in the highest degree improbable that so notable a personal refinement should have remained unknown here, when for three centuries before it had been practised in France. M. Challamel (*History of Fashion in France*) says that glove-makers were selling in the latter part of the thirteenth century " gloves made of kid prepared with violet powder," and in the next century, under the rule of Charles VI. and Charles VII., " violet-scented gloves." These latter were, according to Oliver de la Marche, brought from Spain.

The evident partiality of her dress-loving Majesty for these dainty gloves led to many other similar presents being made to her. Nichols's *Progresses of Queen Elizabeth*, for instance, shows, in 1577-8, gifts—

By the Lady Mary Grey, ij peir of swete gloves with fower dozen buttons of golde, in every one a side perle.

By Lady Mary Sydney, one peir of perfumed gloves with xxiii small buttons of golde, in every one of them a small diamond.

By Petro Lupo, a peire of sweete gloves.

By Josepho Lupo, a peire of swete gloves.

By Cæsar Caliardo, a peire of sweete gloves.

Once patronized effusively by Elizabeth, the fashion took fast hold on the people. Until that time perfumery, although known and used all over the Continent, and held in high repute by all Eastern nations for ages before, did not commend itself to our countrymen. For this neglect they now made ample amends. They scented their handkerchiefs, their cuffs, even their jerkins and their pockets. Housewives became learned in the distillation of sweet waters and in the preparation and compounding of all manner of delicate oils, soaps, powders, essences ; and in the making of balls or pomanders, such as Antolycus carries in his pack (*Winter's Tale*), which were balls filled with perfumes, or dried oranges stuffed with cloves and spices to be carried in the pocket or about the neck, generally as a luxury, but sometimes as a safeguard against infectious disease. Cavendish describes Cardinal Wolsey entering a crowded chamber, "holding in his hand a fair Orange, whereof the meat or substance within was taken out, and filled up again with the part of a sponge, wherein was vinegar and other confections against the pestilent airs ; the which he most commonly smelt unto, passing among the press, or else when he was pestered with many suitors." Pomanders of a more elaborate kind were made of filigree silver work, after the fashion of the vinaigrettes of our grandmothers, and often had separate compartments for holding essences in variety to which medicinal virtue was attributed. "They have in physic," writes Lord Bacon, "use of pomanders and knots of powders for drying of rheums, comforting of the heart, and provoking of sleep." A "cheyne of pomander with buttons of silver betwene" is mentioned in Nichols's *Progresses*, and others are spoken of in the *Privy Purse Expenses of the Princess* (*Mary Archæological Journal*, 1874).

The universality of what must have appeared a flagrant affectation provoked the wrath of Philip Stubbes, in his vigorous *Anatomie of Abuses*. "Is not this," he asks, "a sweet pride, to have civet, musk, sweet powders, fragrant pomanders, odorous perfumes and such like, whereof the smell may be felt and perceived, not only all over the house or place where they be present, but also a stone's cast off almost ? . . . . And in the summer time when flowers be green and fragrant, ye shall not have any gentlewoman almost, no, nor yet any droye (duster or dish-wiper) or pussle (probably puss—conceited puss) in the country but they will carry in their hands nosegays and posies of flowers to smell at, and, which is more, two or three sticked in their breasts, before." In another passage of his book, which is one long and unsparing invective of the practices and dress of his day, he complains that ladies must have their fingers "decked with gold, silver, and precious stones : their wrists with bracelets and armlets of gold and costly jewels ; their hands covered with sweet-washed gloves."

Much of this very conspicuous refinement must be attributed to the example of Elizabeth and her courtiers, particularly of the noble earl who had returned from Tuscany so "Italianated." Before fashion-magazines multiplied, and the restless energy of milliners invented something new or revived something old every week to keep pace with a hunger constantly stimulated by "Our Paris Correspondents," dress had for a time distinct boundary lines of which the queen regnant was generally the arbiter and rule of right. Indeed, until comparatively recent times, our sovereigns were the models of fashion, if not invariably the mould of form, and the changes which follow in the prevalent costume whenever any monarch took to wife a foreign princess are frequently very marked.

> The people vary, too,
> Just as their princes do,

wrote Claudian—a maxim particularly verified in the history of fashion. The imitation which is, proverbially at least, the most sincere of flattery, sometimes put a malformed monarch at his ease by duplicating a too-conspicuous deformity. The Emperor Francis, according to an old chronicler, set the wearing of short hair in 1529, on a journey from Barcelona to Genoa, "cutting it off as a vow for his Passage," or, as our authority artlessly says, "as others have it, for a Pain in his Head." None of his courtiers would presume to wear the time-honoured ringlets in sight of a kingly cropped poll, and so began a barbing and a shearing of the luxuriant locks until every Samson of them fell a victim to the Delilah of Fashion. High shoulders, as Addison humorously shows, were in favour during the reign of Richard the Crookbacked, and when Alexander the Great wore his head involuntarily a little over the left shoulder "not a soul stirred out till he had adjusted his neck-bone; the whole nobility addressed the prince and each other obliquely, and all matters of importance were concerted and carried on in the Macedonian Court with their polls on one side." Is this farcical nonsense? What, have we not in our own day suffered from Alexandra limps and Grecian bends to prove the existence of the folly which will follow any extravagance or imperfection observable in those who sit in high places? Elizabeth began the wearing of ruffs to hide a wen in her neck, or, as some declare, to cover a yellow neck. Forthwith ruffs blossomed round every throat, and spread until grave citizens stood at city gates to cut down all exceeding an ell in breadth. High-heeled shoes were revived by Madame Pompadour to correct her short stature, and then ladies of all heights tottered on pedal pedestals; cravats were taken up by a lady who, like the Saxon Princess Edith, was swan-necked, only much more so; and short skirts were adopted by a leader of fashion who knew that in ankles at least she had a decided advantage over her compeers.

To increase the hold of the new refinement of perfumery

upon the people, a similar movement was fostered in France by Catherine de Medicis. Even among the middle classes, sexes there vied with each other in the lavish employment of scent, and ingenuity was sadly exercised to find new means of declaring a willing servitude to the demand of the day. All apparel was perfumed ; hair and shoes and fans gave out a sweet-smelling savour, and all kinds of jewellery contained cavities filled with strong essences. Nicolas de Montau, in his *Miroir des Francois*, reproved ladies for "making use of every perfume—cordial, civet, musk, ambergris, and other precious aromatic substances— for perfuming their dresses and linen, nay, their whole bodies." Perfumed gloves were not the least conspicuous of these toilet accessories.

An ordinary method of imparting the scent by means of animal essences was to mix the substance with some kind of oil and then rub it into the glove. In other instances advantage was taken of the ready absorption by fatty substances of the odours of flowers, which is the basis of all perfumery ; some manner of pomatum, as we may call it, was prepared and then plastered all over the inner surface of the glove. All manner of fragrant herbs and drugs were used, singly and in combination, to give gloves this desirable quality. Two simple and cleanly old recipes, within the scope of any good housewife, directed to this end :—" Put into angelica water and rose water the powder of cloves, ambergris, musk, and lignum aloes ; benjamin* and carduus aromaticus. Boil these till half be consumed, then strain it and put your gloves therein. Hang them in the sun to dry, and turn them

---

* Benjamin, benzoin, " The aromatical gumme called benjamin or benzoin.— COTGRAVE.

often.   Do this three times : wetting and drying them
again.   Or, wet your gloves in rose water and hang them
up till almost dry, then grind half an ounce of benjamin
with oil of almonds, and rub it on the gloves till it be
almost dried in.   Then grind twenty grains each of
ambergris and musk with the oil of almonds, and rub it
on the gloves.   Then hang them up to dry, or let them
dry in your bosom, and so, after, use them at your
pleasure" (MARK).   In Andrew Borde's *Regyment* his
directions on *Sleep, Rising, and Dress (Early Eng. Text
Soc.)*, including very minute and most amusing directions
on all these heads, enjoins that in summer you should
"kepe your neck and face from the sonne ; vse to wear
gloues made of goote skyn perfumed with Amber de
grece."

Other perfumed gloves were at times given to Elizabeth.
"Three Italians came unto the queen and presented her
eache of them a payre of sweete gloves" (Nichol's
Progresses), and they became so universal that Warton says
that they were often given at the Universities to college
tenants as well as to guests of distinction, so that an item
in the Bursar's books of Trinity College, Oxford, is
"pro fumigatis chirothecis." They are mentioned by
Shakespeare.   Antolycus sings of—

Gloves as sweet as damask roses,

and in *Much Ado about Nothing* Hero says to Beatrice—

These gloves the Count sent me, they are an excellent perfume.

To James I., in 1606, there were presented " By
William Huggins, one payre of perfumed gloves, the cuffs

laced with point-lace laces of Venice gold ; " and "two payre of plaine perfumed gloves ; " " Also, by the King's musicians, eche of them one payre of plain perfumed gloves." Spain, as we have shown, early obtained a reputation for scented gloves, and her trade, both in embroidered and perfumed gloves, as well as in skins already scented but not cut up, is said to have existed for several centuries, lasting at any rate until the time of Cervantes, for the immortal knight says to Sancho Panza, " This you will not deny, Sancho, that when you were so near her, your nostrils were regaled by a Sabæan odour—an aromatic fragrance—a delicious sensation for which there is no name : I mean a scent such as fills the shop of some curious glover." Howell, a seventeenth-century writer, says, too, " In the Grand Cairo, when the wind is southward, they say the air is as sweet as a perfumed Spanish glove ; " and Ben Jonson, in his comedy, *The New Inn*, includes among the dress of a beau " My gloves, the natives of Madrid."

The fragrance imparted to the Spanish gloves was of a very enduring character. The French, on the contrary, trusting more to the distilled oils of natural flowers, failed in giving permanency to the perfumery of their gloves and other articles of attire.

In the seventeenth century there appears to have been a general revival of effeminacy and ultra refinement. Evelyn, best known by his diary, but author besides of several other works—scientific and otherwise—published, in 1690, a playful satire, entitled *Mundus Muliebris ; or the Ladies' Dressing-room Unlock'd and her Toillette Spread*, which is now a valuable book of reference for the costume of the period. Among the requisites of the wardrobe are reckoned "Gloves trimm'd and lac'd as fine as Nell's"—a familiar allusion to Nell Gwynne, mistress of Charles II. :—

> Twelve dozen Martial whole and half,

that is, long and short gloves, "so called from the Frenchman's name pretending to make them better than all others " (*Ladies' Dict.* 1694).

> Of jonquil, tuberose (don't laugh)
> Frangipan, orange, violett,
> Narcissus, jassemin, ambrett.

Frangipan gloves also were named after the Marquis Frangipani, Maréchal des Armées of Louis XIII., he having invented a new method of imparting perfume to gloves. Menage, in his *Origini della Lingua Italiana*, 1685, thus notices the Marquis and his invention : " Da uno di que Signori Frangipani (l'abbiam veduto qui in Parigi) furono chiamati certi guanti porfumati *Guanti di Frangipani*" (*N. & Q.*). In France they wore at this time gloves "à l'occasion," "à la nécessité," "à la cadenet," "à la Phyllis," "à la Frangipani," and gloves "à la Neroli;" the latter again perpetuating the name of a princess who had the honour of inventing a new and exquisite scent. Perfumed gloves were then almost universally worn, and in the Letters Patent granted by Louis XIV. to the French company, in 1656, the glovers were styled " Marchands Maîtres Gantiers Parfeumeurs," and were privileged to make and sell gloves, mittens, and skins used in making gloves, to perfume gloves, and sell all manner of perfumes.

In Venice particularly the *furore* for perfumery ran

high. Delicacy reached an extreme pitch. Ladies, emulating the dainty dames of Rome when the canker of indolence and luxury were eating away the foundations of a noble empire, launched into unbridled extravagance and dissipation. Stringent sumptuary laws failed to put bounds on excess in dress, and display defied control. Into their baths fair Venetians would throw musk, amber, aloes, myrrh, cedar leaves, lavender, mint, and other fragrant herbs and spices ; the toilet table was crowded with cosmetics, and the art of Madame Rachel flourished exceedingly. The love of dress in Venice is traced by historians to the year 1071, when Domenico Seivo, the Doge, married a Constantine Ducas, who brought with her Eastern customs which first astonished and then corrupted the Venetian ladies, who had, until that time, been simple in their attire, and comparatively ignorant of luxuries. To this grand dame we may, to all seeming, attribute the introduction of perfumed gloves into Europe. She might have been the prototype of Shakespeare's Cleopatra, this woman who made the winds " love-sick with perfume," who washed in scented water every morning, who covered her body with aromatic oils and unguents, and bathed her face in dew which she made a multitude of slaves collect for her. She is strongly condemned by chroniclers for her perfumed gloves, and scorned, too, for wearing silken vests, no less than for using a gold stick to convey her food to her mouth in place of following the ancient usage of employing fingers and napkins, a practice which, through our table napery, still has some connection with the present. Happily, however, we have for some three centuries past ceased to associate forks with foppishness, as did Ben Jonson and

Beaumont and Fletcher, with other of their contemporaries; and it would cause some sensation if a learned divine now preached a sermon against the use of forks, as did one in James the First's reign, declaring it to be "an insult on Providence not to touch one's meat with one's fingers."

Gloves made of Russia leather, for the sake of the peculiar fragrance for which some people have so great a partiality, have been in wear during the few past years, as well as gloves scented with cedar wood and violets. Indeed, the practice with the French of perfuming gloves has never wholly died out, and within the present century they have worn them scented with myrtle, although even French domination in fashions could hardly again establish for perfumed gloves such a supremacy as was in former times allowed them.

CHAPTER IX

Chicken Gloves.

ONTINUING the enumeration of the various gloves considered necessary to complete the amorous armoury of a seventeenth-century belle, Evelyn includes—

> Some of chicken skin for night,
> To keep her hands plump, soft, and white.
> *—Mundus Muliebris.*

The practice of wearing gloves at ' night to impart particular delicacy to the skin was common to gentlemen as well as ladies, and was followed as late as the reign of George III. To even greater lengths did some votaries of fashion go ; it was but a mild measure to lard the face over at night, nothing extraordinary to wear gloves lined with unguents, or to cover the face with a mask plastered inside with a perfumed pomade to preserve the complexion. Some steeped slices of raw veal for some hours in milk and laid them on the face. Young and tender beauties bathed in milk ; beauties who were no longer young, and far from tender, bathed in wine or some other astringent.

Gloves of chicken skin appear to have been held to have particular virtue in giving the wearer a hand to be proud of, and, although no trace can be found of such a manufacture, it seems undeniable that chicken skins have been used for the purpose. Indeed, since we know that old fans are often covered with chicken skin, to be painted upon, there is no reason to doubt that the material may have been used for gloves. Too intimate an acquaintance at dinner with a very mature fowl affords ready proof at any time that the skin is at least available for such a purpose. In the *New Bath Guide*, a scandalous publication, not out of place in the time of Fielding and Smollett, but hardly suitable for reproduction in our day, there is given a fanciful description of the dress of the period, which includes an imaginative account of the introduction of chicken gloves :—

> Come, but don't forget the gloves
> Which, with all the smiling loves,
> Venus caught young Cupid picking
> From the tender breast of chicken ;
> Little chicken, worthier far,
> Than the birds of Juno's car,
> Soft as Cytherea's dove,
> Let thy skin my skin improve ;
> Thou by night shalt grace my arm,
> And by day shalt teach to charm.

Ben Jonson, in his play of *The Alchemist*, 1610, introduces gloves of this description as a superlative refinement, but it is impossible to determine whether chicken gloves were made at this early date or whether the poet was inspired by a fine imagination or spirit of prophecy. Sir Epicure Mammon, addressing Surly in Lovewit's house, says—

> My shirts
> I'll have of taffeta-sarcenet, soft and light
> As cobwebs ; and for all my other raiment

It shall be such as might provoke the Persian
Were he to teach the world to plot anew.
My gloves of fishes' and birds' skins, perfumed
With gums of Paradise and Eastern air.

It was truly entertaining to find a New York paper, in quoting this passage as an instance of supreme Sybaritism, remark upon it, " Dr. Johnson (!) had an idea of nice furnishing goods."

It is certain that if gloves were at any time largely manufactured from chicken skins they were soon superseded by other thin and fine materials, so that chicken gloves became as great a misnomer as the Speaker of the House of Commons—the most silent of all the active members—or Irish stew, an unknown dish in Ireland. The seat of the manufacture was at Limerick, but it was also carried on at Waterford and Dublin. The gloves were there made from the skins of unborn calves, either taken from cows that died (when the skins were technically known as "morts") or from parturient cows purposely slaughtered, a barbarous practice once common in Ireland. These latter skins were called in the market " slinks." As may be supposed, these gloves were of very fine texture, so much so that they could be enclosed in a walnut shell, and were thus often·shown in shop windows. A pair of them were included in Ralph Thoresby's museum —" A pair of gloves so delicately thin that, though they will fit a large hand, are folded up and enclosed in a gilded walnut's shells." It was this quality that gave " Limericks," as the gloves were collectively called, their extensive reputation.

Gloves, if not called "chicken," yet of equally fine material, were made in Scotland. The Incorporation of

H

Glovers of Perth—once a powerful and wealthy craft—
had in a coat of arms sometimes used, although not the
recognized cognizance of the trade, *five balls or nut-
shells* on a branch placed between a pair of gloves
"displayed" on a shield, and beneath those a pair of
large glove-cutting shears, the skinners' paring iron and
a pair of glove-sticks. A stag and goat, both "rampant,"
supported the shield on which these trade implements and
emblems were borne, with a ram "passant" as the crest.
A former Deacon of the Incorporation was of opinion
that this, differing from the recognized arms of the calling,
was, even though found on the "calling's seats" in Perth
churches and emblazoned on some of their Minute Books
and other records, due to the imagination of the painter
rather than to be accepted as of historical value. But in
an old picture of St. Bartholomew formerly hung in the
Perth Glover's Hall there was in the corner, again, a bunch
of these nut-shells or balls, and the Deacon on this
remarks that "the fine balls or nut-shells were used for
the purpose of containing specimens of the manufacture
of certain descriptions of remarkably fine gloves in
ancient times, for they were made of such fine materials
that they were folded in pairs and enclosed in these
nut-shells, which were often sent as presents by the
cavaliers of olden times to the fair sex as tokens of
affection and love."

Among the innumerable schemes advocated for the
improvement of the unfortunate condition of Ireland—to
stay the perennial discontent which follows so hard upon a
population condemned to perpetual poverty for want of
proper employment—the resuscitation of the glove
industry has not been lost sight of. Irish gloves, it is

urged, once had wide fame: could they not be re-established and become a profitable industry, carrying comfort and contentment to many a village home where broken-spirited men brood over their hopeless prospects and become ready prey to unscrupulous agitators? It has been Ireland's misfortune that none of the domestic avocations established there from time to time by her truest patriots have acquired the element of permanence. Fashion ruined the sewed-muslin trade, which at one time gave a fictitious prosperity to the Northern and Western counties. Sericulture and straw-plaiting were both started with apparently every hope of success. Knit hosiery was once largely practised, as well as hand-loom weaving—both industries which had in them, to all seeming, the very elements of hope, with the additional advantage of utilizing common industrial knowledge. Lace-making was another occupation which seemed to meet a known need, suiting the dexterous and pliant fingers of the peasants, and employing the cheapest of labour in the production of a tediously-produced, but costly material. None of these efforts succeeded ; not from lack of capital or the absence of coal—two standard excuses for Ireland's failures. But whether from lack of application, or through dispo_ sitions too quickly spoilt by success ; whether the patrons had not sufficient commercial knowledge, or lost heart or interest ; from some reason or another all these enterprises have either failed or have so feeble an existence as to be altogether unworthy consideration, except to add another chapter to the pitiful history of a people not inferior to any other in intelligence, strength, or abilities.

The decadence of the Irish glove trade has been attributed to French competition, not in general only, but by.

direct effort. The Irish workmen, says one apologist, are alone to blame. " While the trade was still flourishing the French workmen became alive to the superiority of Irish kid skins and the Irish method of dressing them, and came to our country to learn. Having gained all the information they sought they returned, taking with them several Irish workers, and with all the skins they could buy up in the Irish market. Such a proceeding, though in nowise reprehensible on their part, produced a comparative scarcity in the home market, and the skill of the Irish workmen, aided by the use of machinery (there was not any machinery in Ireland), enabled French manufacturers to produce excellent gloves in large quantities. The home makers raised their prices, which the trade refused to accept, and abandoned the Irish for the French manufacturers."

Hull, in his *History of the Glove Trade*, assigns a high place to the glove industry in Ireland ; it gave, he says, " extensive employment to many thousands of people in Ireland ; " and he further states that " the glove trade in Ireland not only occupied many thousands directly in the trade, but it gave occupation to an immense number of persons who went all over Ireland collecting the skins for the gloves, and on an average one million skins were collected and consumed." These statements require to be taken with more than one grain of salt. The latter is obviously exaggerated ; and we may conclude that all the figures are over-estimated, and give the glove trade of Ireland an entirely fictitious importance, unless the best writers on Irish industries are culpably silent on the subject. It is rare—and in this particular the writer can claim an extensive acquaintance with works of the kind—that gloves are mentioned at all in any account of Ireland's

manufactures. It is absolutely untrue that any such prosperity attended the glove trade in Ireland within this
century, and, indeed, Hull admits that it had in his time
(1834) utterly decayed. To go farther back, Wakefield
(*Account of Ireland*, 1812) says, "Gloves are manufactured
in Ireland, but not to any great extent." Arthur Young's
Tour in Ireland, 1780, does not mention gloves at all, and
Lord Sheffield's *Observations on the Manufactures of Ireland*,
1785—treating of the products of the country *seriatim*—
does not include gloves.

It is fair, in considering the matter, to note that Lord
Sheffield shows from the Custom House books a declared
export, in 1783, of 22,510 dozen of calves' skins, sent
almost entirely to Scotland and Ireland. These, we are
free to admit, were possibly used for the greater part in the
manufacture of gloves. It is further asserted that "great
frauds are committed in the entry of hides, and particularly
of calves' skins outwards; there is a duty on the export,
and it is certain that the quantity exported exceeds greatly
the quantities entered in the Custom House books."

Still, allowing this illicit traffic to be equal to the legiti-
mate trade, presuming that some of these skins were used
for producing light leather articles and were wholly used up
in glove-making, adding another like quantity to be made
up into Irish gloves, and there yet remains a large balance
to make up an average annual produce of a million skins,
and equal difficulty in imagining the services of an "immense" number of persons to be required in collecting
them. In this year (1783) there were no gloves exported
from Ireland to Scotland or England, the principal channels
of trade, and by reason of the Navigation Act the only
considerable channels of export. To Newfoundland, Ireland

sent 48 pairs of gloves, and to Nova Scotia 1,014 pairs. From England were brought in 743 pairs, at an estimated value of 3s. per pair. A large trade, too, was done in French gloves, although the figures for 1783 are not shown. But, in 1765, 5,747 pairs of French gloves were imported into Ireland, 5,030 pairs in 1766, 12,726 pairs in 1775, and 4,176 pairs in 1776, showing a large and generally stable trade.

It would not be possible to cite higher authorities on Irish trade than those quoted, but it is remarkable that of the large number of tracts on Irish trade all are silent as to this great glove traffic, and it is most probable that even if the flood-mark of Irish prosperity in glove manufacture were again reached the country would hardly be sensible of it. The idea of making the industry a profitable or extensive one in these days must be pronounced altogether hopeless. It is very easy to advocate that a factory shall be established in every town of importance, and Irishwomen induced to make each successful by patronizing only home manufactures ; but Ireland has, unfortunately, no possibility of becoming self-supporting, and, in this respect, no hope of being able to encounter the competition of France or England.

The only remaining connection of Ireland with glove manufacture is the existence of a solitary factory at Cork, and an inconsiderable trade in kid-skins, which are annually collected, chiefly by French agents.

CHAPTER X.

Gloves in Common Wear.

LONG before gloves were made here an evidence of refinement or mark of rank, they were known and widely worn upon the Continent. Articles made from leather, harness, belts, and the like, as well as gloves, were from a very early date manufactures in Germany. Gloves for men left the fingers free, as mittens do, but had a thumb-stall. Those worn by ladies were of finer material and workmanship, often adorned at the back with a number of precious stones or a single jewel of price, after the manner of those worn by princes and prelates at a later period, and were, in many instances, made so long as to reach half way up the fore-arm or to the elbow (YEATS : *Technical History of Commerce*). Fur gloves were by no means unusual in the time of Charlemagne, and the skin most generally made up was that of the otter, which old Izaak Walton so strongly recommends as " the best fortification for your hands that can be thought on against wet weather." Monks, we learn, also wore, in winter, gloves of sheep-skin for the sake of warmth. Whether they were allowed the privilege unrestricted and without scruple is not stated. Probably not. It was not often that the servants of the early church were allowed to consider the frailty of the flesh, or indulge themselves in comfort without special dispensation. Only after a formal application to the Holy Father, and with his special consent, were the nuns of Montmartre permitted to wear fur-lined boots, and a like privilege had to be sought for monks to wear *super-pellis*—covering furs—above their vestments when reciting the offices at dead of night and in inclement weather, and so inaugurate the wearing of surplices.

We went on in benighted ignorance, careless of culture, destitute of gloves. Saxons succeeded Britons, the Danes came and went, and the Normans came, and did not go, before gloves had a recognized place in our national costume. Then the foppery of young Norman nobles, as Ordericus Vitalis complains, led them to cover their hands with gloves too long and wide for doing anything useful. They had them reaching nearly to the elbows, and ornamented at the tops with embroidery. (STRUTT). It is a remarkable fact that the use of gloves was at this time and for long afterwards confined to men. Man in those days was not the sombre-hued individual the nineteenth century has made him, given over to what the statutes of Oxford call "subfusk" apparel—the only male in all creation soberer in plumage than his mate. He asserted then the prerogative of his sex, and clothed himself in rich array, making himself conspicuous by gaudy raiment and "bravery." Woman did not then so much make fashions as follow them. That she only ventured to wear gloves after their use had become very general among the other sex is only one of many like instances familiar to students of costume. It has been urged that this anomaly, as it appears in our day, was by no means one of the mediæval disabilities of women, but that the sex then, as ever, first wilful and afterwards firm, only did just what best pleased themselves, and that we must not wonder at or argue from a caprice so truly feminine. So, too, in mitigation of this astonishing truth as to gloves, we are bid to remember that ladies of the olden time covered their fingers with as many rings as could be carried on them, like unto their sisters of Hindostan, but without wearing them on toes also. This is true enough ; and, further, ordinary sleeves were worn of such length that they could be let down at will to cover the hand and give all desirable protection, a fashion followed long before by both sexes among the Saxons to such an extent that the sleeves, when not let fall, appear in thick folds all over the arm. Still, the absence of gloves in Norman female dress is so conspicuous, even when they were reckoned marks of refinement among men and made vehicles of display, that something more than this is necessary to show women free from control in their dress. It is more to the purpose that mittens were frequently worn. The term "mittens" has been before now made to denote entire gloves (RAY : HALLIWELL); but the mittens of the fourteenth and following century are distinct enough. Gough (*Sepulchral Antiquities*) shows several varieties on effigies, illustrating each by examples. The mittens were sometimes made of one piece with the sleeve, and so appear on the figures of lady Harsicke (1365) and Stapleton (1384), and on a lady at Easton, Suffolk, of the fifteenth century. They come from under the sleeve of the wife of Henry of Nottingham, at Holm, Norfolk, in the reign of Henry IV., and of Cecilia de Kerdeston (1391). They are buttoned on the Elstow abbess, and they seem to be distinct from the sleeve on Maud de Cobham, Lady Burgate (1419); on Frances Poynings, in St. Helen's, Bishopsgate; and on the abbess at Goring. Mittens, during the Plantagenet period, were common to all classes. They are mentioned in *Piers Plowman's Crede*, and by Chaucer, and a four-teenth-century shepherd, called to court (HARTSHORNE: *Metrical Tales*), was careful not to forget his "mytans," so that when directed to leave his staff and mittens before entering the court he will not part from them :—

> My staff shall never go from me,
> I will kepe it in my hande ;
> Nor my mytens gets no man,
> While that I them keep can.

It would seem from this universal wearing of mittens that gloves would be superfluous, and we might so conclude, were it not that men continued to wear them both in civil costume and in the habiliments of war.

Coming to the fourteenth century, we meet with some traces of the adoption of gloves by ladies, but only at first, as it would appear, by ladies of rank. Gough says that gloves were not worn at all by women before the Reformation ; but the statement, although substantially accurate, is not absolutely so. Gower—moral John, as Chaucer calls him—in his *Confessio Amantis* writes,—

> But er thei go, some aduantage
> There will they haue, and some pillage
> Of goodly wordes, or of beheste
> Or elles thei take at leste,
> Out of hir honde a rynge or gloue.

And Chaucer himself, in his translation of *The Romaunt of the Rose*, gives "Idleness" a pair of gloves—

> And for to keep her honds fayre,
> Of gloves white she had a payre.

Fairholt engraves (unfortunately without specifying the source from which it is taken) the figure of a lady "from a manuscript of the fourteenth century, holding in her hand a glove." In Walker's *Historical Essay on the Dress of the Ancient and Modern Irish* is a plate representing the figure of a lady bearing a glove in her hand, taken from a tomb of the fourteenth century, unearthed from a heap of rubbish in the Priory of Athassel. So, too, in the monastery of St. Saviour's, at Ross, in the county of Wexford, is a monumental figure of Rose Macrue, at whose expense the town was walled about in 1310. This figure also holds a glove in one hand. Lysons testifies to having seen "a pair which belonged to the Duchess of Exeter, sister to Edward IV. (?)," and he speaks of them as being richly embroidered, very much resembling "the wedding gloves of Mrs. Hampden, wife of the celebrated patriot, which are now in the Earl of Oxford's collection at Strawberry Hill." These cases collectively point, at any rate, to the occasional wearing of gloves by ladies of high rank before the Reformation, although there is no doubt that the fashion did not become general before that time.

As bearing indirectly upon the question as to when ladies first wore gloves, but more for its intrinsic interest, space must be spared for a synopsis given by Warton of a fourteenth-century metrical fiction, entitled *The Romance of Syr Degore and the Fierce Dragon*, in which a pair of gloves is made the pivot of its plot. A king's daughter, extremely beautiful, is solicited in marriage by numerous potentates of various kingdoms. The king, her father, vows, that of all these suitors, that champion alone shall win his daughter who can unhorse him in a tournament.

This they all attempt, but in vain. The king every year assisted at an anniversary mass for the soul of his deceased queen, who was interred in an abbey some distance from his castle. In the journey thither the princess strays from her damsels in a solitary forest. She is discovered by a knight in rich armour. In his company she proves as frail as fair, and at parting he gives her a pair of gloves, which will fit no hands but her own, and a sword having no point, charging her to keep them safely. Ultimately, the princess, regaining her father's castle, is secretly delivered there of a male child. Soon after this event, the princess, having carefully placed the child in a cradle, with twenty pounds in gold, ten pounds in silver, the gloves given her by the strange knight, and a letter, consigns him to one of her maidens, who carries him by night, and leaves him in a wood near a hermitage, which she discerned by the light of the moon. The hermit, in the morning, discovers the child, reads the letter, by which it appears that the gloves will fit no lady but the boy's mother, educates him till he is twenty years of age, and at parting gives him the gloves found with him in the cradle, telling him they will fit no lady but his own mother. The youth, who is called Degore, sets forward to seek adventures, and saves an earl from a terrible dragon, which he kills. The earl invites him to his palace, dubs him a knight, gives him a horse and armour, and offers him half his territory. Sir Degore refuses to accept this offer unless the gloves, which he had received from his foster-father, the hermit, will fit any lady of the earl's court; but upon trial of the ladies, including the earl's daughter, the gloves will fit none of them. He therefore takes leave of the earl; proceeds on his adventures; in due course meets with a long train of knights, and is informed that they are going to tourney with the king of England, who had promised his daughter to that knight who could conquer him in single combat. They tell him of the many barons and earls whom the king had foiled in several trials. Sir Degore, however, enters the lists, overthrows the king, and obtains the princess. As the knight is a perfect stranger she submits to her father's commands with much reluctance. Sir Degore marries her, but in the midst of the solemnities recollects the gloves which the hermit had given him, and proposes to make an experiment with them on the hands of his bride. The princess, on seeing the gloves, changes colour, claims them for her own, and draws them on with the greatest ease. She declares herself to Sir Degore as his mother, and narrates the circumstances attending his birth, telling him of the pointless sword the knight his father had given to her to be delivered to none other than their child. Sir Degore contemplating with wonder the length and breadth of the weapon, is seized with a desire to find his father. He sets forward on his search, and on his way enters a castle, where he is entertained at supper by fifteen beautiful damsels. The lady of the castle endeavours by various artifices to divert him from his quest, and even engages him to encounter a giant on her behalf. Continuing steadfast in his aim, Sir Degore goes on his way, and in a forest next meets a knight richly accoutred, who roughly demands the reason why the stranger had presumed to enter his domain without permission. A combat ensues. In the midst of the contest, the combatants being both unhorsed, the knight, observing the sword of his adversary to be unusually long and broad and pointless, begs a truce for a moment; and then fits the sword with a point which had always been in his possession, and which had formerly been broken off in an encounter with a giant. So father and son are discovered to each other. They return to England together, and Sir Degore's father is married to the princess his mother.

This story in its opportunities of moving incidents of combats between knights, and fights with dragons, of descriptions of love scenes and magnificent tourneys, would suit well the humour of our ancestors. It is more to modern tastes to turn to mention made of gloves in a poem by Robert Henryson, a fifteenth-century schoolmaster. This, *The Garmond of Fair Ladies*, sets forth the symbolic dress of a true woman: hood of honour; " sark" of chastity, made up with shame and dread; kirtle of constancy, laced with lawful love through eyelets of continuance; gown of goodliness, furred with fine manners and edged with pleasure; belt of benignity; mantle of humility; tippet of truth; sleeves of hope; shoes of security; hose of honesty, and

> Gluvis of the gud govirnance
> To hyd hir fingearis fair.

(FAIRHOLT : *Satirical Songs and Poems on Costume* [Percy Society].)

Although allegorical garments are at least as ancient as St. Paul, these quaint and beautiful verses have a curious parallel in *Le Triomphe des Dames*, written by a French poet, Olivier de la Marche, in 1464. In this, ladies are recommended to wear slippers of humility, shoes of diligence, stockings of perseverance, garters of firm purpose, a cotte of chastity, a waistband of magnanimity, a ring of faith, a hood of hope, and to carry a purse of liberality.

To return to the chronological sequence of this chapter;

the *History of the Exchequer* (MADOX) shows a payment in the time of John of £28 18s. 2d. for "c. and xl. paribus cirotecarum," but these were probably gauntlets, more particularly in view of the price paid for them. About this time, gloves of coarse warm materials, and without separate fingers, made part of the costume of the commonalty, the higher classes, both in church and state, wearing them of rich materials, and jewelled at the back. Luxury in dress was then provoking popular censure. "A Song upon the Tailors" (*Harl. MSS.* 978), printed and translated by Thomas Wright in his *Political Songs* (*Cam. Soc.*), is only one of many rebukes of excess in apparel, which probably had as little effect then as the phillippics of physicians and the condemnation of the press have in our own day. The song, addressing the tailors, exalts them as gods, able to make old garments new. "The cloth while fresh and new, is made either a cape or a mantle ; but, in order of time, first it is a cape, after a little space this is transformed into the other ; thus ye change bodies. When it becomes old, the collar is cut off ; when deprived of its collar, it is made a mantle ; thus, in the manner of Proteus, are garments changed. When at length winter returns, many engraft immediately upon the cape a capuce ; then it is squared ; after being squared it is rounded, and so it becomes an amice. If there remain any morsels of the cloth or skin which is cut, it does not want a use ; of these are made gloves.* This is the general manner they all make one robe out of another—English, Germans, French and Normans, with scarcely an exception.''

In the paintings formerly on the walls of the old Palace

---

* What is here rendered " gloves " is in the original " manuthecæ." This may mean gloves, but it may also mean, and more probably should be, *sleeves*.

of Westminster, attributable to the close of the thirteenth century, the figure of Antiochus bears long, tight-fitting white gloves, reaching half-way up the forearm, with a broad stripe of gold embroidery down the back from the top to the knuckles. Planchè engraves a detail of this embellishment. Illuminations of this period show that it was a common practice then to carry gloves in the hands, or to thrust them through the girdle, an anomalous practice showing gloves neither entirely useful nor alto-gether ornamental, which now and again became common. The ladies and damsels of the court of Catherine de Medicis are so shown, carrying gloves in their hands, and one of the characters in an emblematic show in the Lord Mayor's pageant of 1664 wore "a gold girdle, and gloves hung thereon."

In the sixteenth century we meet with frequent mention of gloves, including those knitted. Knit gloves carry us back at once into the burden and heat of a fierce literary fight. There can be no doubt now that the Poems of Rowley were, however ingenious and gifted, still only literary frauds, and that—

> The marvellous boy,
> The sleepless soul that perished in his pride,

was, from one point of view, on a level with Ireland, the Shakesperian forger. But around the authenticity of the verse of the fictitious Bristol monk once raged as fierce and prolonged a controversy as that provoked by the poems of Ossian. The battle began over a passage in the "Mynstrelle's Songe, bie Syr Thybbot Gorges," in the *Tragycal Enterlude of Ælla :—*

> As Elynour bie the greene lesselle was syltinge,
> As from the sone's hete she harried,
> She sayde, as herr whytte hondes whytte hosen were knyttinge,
> Whatte pleasure ytt ys to be married.

It was contended that Rowley, a fifteenth-century monk, as Chatterton represented him, could have known nothing of knitting, an art alleged not to have been invented until the following century. Those who would like to know a little of the amenities of a quarrel of this order should turn to the *Gentleman's Magazine* for 1781 and 1782. The generally accepted, but erroneous, account of the introduction of knitting is taken from Stow, who says, "In the year 1564, William Rider, being an apprentice with Master Thomas Burdett, at the Bridge foot, over against Saint Magnus Church, chanced to see a paire of knit worsted stockings in the lodging of an Italian merchant that came from Mantua ; borrowed those stockings, and caused other stockings to be made from them, and these were the first worsted stockings made in England." Not the first knit stockings known here, though, nor the beginning of the art of knitting. Extracts under the years 1533 and 1538, from a *Household Book of Sir Thomas L'Estrange*, of Hunstanton, prove that "Knytt hose" were then known and purchased here. Henry VIII. had in his wardrobe white silk and gold hose, knit, and six pairs of black silk hose, knit. Further, poor Chatterton might have found in this respect a complete answer to his opponents in an Act of Parliament of 1488, which mentions knitted woollen caps. A subsequent statute of 1551, mentions "knitte hose, knitte peticotes, knitte gloves, and knitte sleves," and indeed Stow himself says that, prior to the time of Henry VIII., "the English vsed to ride and goe winter and summer in knit capps." As to the higher antiquity of knitting, it has always been believed that the ability to make needle-points take the place of fingers was known to ancient nations, and it has since been proved that the

I

Egyptians could knit stockings, of which a specimen, made of fine wool, has been on view in the Louvre collection. They anticipated, too, a modern improvement, which is made part of the platform of "hygienic dress" reformers, for these stockings are not simply feet-bags, but are divided at the point.   They are not, however, made as it is said we should wear them, with separate receptacles for each toe, but had simply one division, so as to admit of the sandal-strap being held secure and properly fastened.

Knitted gloves of silk were worn in France during the reign of Henri III., and among the gifts to Queen Mary on New Year's Day, 1536, were "two peire of working gloves of silke, knit," from "the lady Grey, the Lorde John Graye's wif."   In a list of "such articles of trade as may be imported into England from the Low Countries," dated 1563, and preserved among the Cecil papers at Hatfield House, there is the following item : "Gloves knytt of sylke."   The various articles in this list are divided into "necessarye" and "superfluouse," the gloves being classed under the latter head.   Clericals took to the wearing of them.   In the *Inventory of Whalley Abbey* (28 Henry VIII.), appear "A pair of knette gloves with a roose of gold embroydered, sett with perls and ij small safours in eyther of them."

Ordinary gloves also occur frequently in sixteenth-century records.   Turning to the *Calendar of State Papers*, which is so invaluable to students of the bye-ways of history, we find in 1549 William Rogers sending to Sir T. Smith, by one Mr. Holinshed, "six pairs of double gloves;" and in 1561 Mabell Forteskew, writing to Fr. Yaxley (whom she addresses as "good Governor"), urges him to ascertain the cause of her mother's anger, and "requests some gloves." What double gloves were we have now no means of showing, and a still more inexplicable entry of 1580 shows that gloves might be made in some degree evidence of the age, disposition, and ghostly condition of the wearer :—

Owen Lloyd to Wm. Pryse,—Desires him to send 16 pair of Oxford gloves of the finest, of 5 or 6 groats a pair, of double Chevrell, 6 for women, 6 for men, and 4 for very ancient and grave men, spiritual.

These Oxford gloves are found mentioned in an inventory of the effects of James Backhouse, of Kirby Lonsdale, taken in 1578—

ij pare of Oxford gloves ijs. iiijd.

LADY'S GLOVES, SIXTEENTH CENTURY.

(In the possession of Rev. Walter Sneyd.)

The conjecture has been made that these were gloves worn after the manner of the Earl of Oxford, whose connection with glove history has already been set forth. This question is determined by a letter from George Evelyn, brother to the diarist, to his father Richard, at Wotton, 26 September, 1636, giving an account of the visit of the King and Queen to the University.

"I know you have longe desired to heere of my well-faire and the totall series of his Majesty's entertainment whilst (hee) was fixed in the center of our academie. The Archbiship oᵣ Lᵈ Chauncelour (Laud) and many Bishops, Doctor Bayley oᵣ Vice Chauncelour wᵗʰ the rest of the Doctors of the University, together wᵗʰ the Maior of the City, and his brethren, rode out in state to meet his Majesty. The Bishops in their Pontificall robes, the Doctors in their scarlet gowns and their black capps (being the habite of the University), the Maior and Aldermen in their scarlet gowns, and 60 other townsmen all in blacke satin doubletts and in old fashion Jacketts. At the appropinquatio of yᵉ King, after the Beedles Stafes were delivered up to His Majesty in token yᵗ they yealded up all their autority to him, the Vice Chauncelor spooke a speech to the King, and presented him with a Bible in the University's behalfe, the Queene wᵗʰ Camden's Britannia in English, and the Prince elect (as I toke it) wᵗʰ Croke's Politicks. All of them wᵗʰ gloves (because Oxford is famous for gloves)."

The Chevrell gloves are famous to a proverb, familiar to readers of Shakespeare. In the third act of *Twelfth Night*, the Clown says to Viola, in a word skirmish—

"A sentence is but a cheveril glove to a good wit : how quickly the wrong side may be turned outward ! "

These gloves were in common wear in the sixteenth century. In the old play of *The Cobbler's Prophesie*, Venus, speaking to her man Nicholas Newfangle, says, in respect to the fickle vanity of women—

"To-day her own hair best becomes, which yellow is as gold ;
A periwig's better for to-morrow, blacker to behold ;
To-day in pumps and cheverill gloves, to walke she will be bolde ;
To-morrow cuffes and countenance for fear of catching cold."

"Cheverill," from the French *chevre*, a goat, denotes kid leather, something more pliable than the coarse skins until this time in use, so that it came to be said of an unscrupulous man, " He has a cheveril conscience "—one that would stretch easily. Palsgrave has, " Cheverell leather ; cheverotin ;" and in the 25th *Coventry Mystery*, " Two dozen points (tagged laces) of "cheverelle" are mentioned. In Sloane's MS. 73, fol. 20, is a curious direction "for to make cheverel leather of parchemyne" (parchment), by means of a solution of alum mixed with yolks of eggs, with flour ; and also to make " whit (white) cheverell, reed (red) cheverell," the colour being imparted to it by a compound of brazil (PLANCHE). " Chevrette " gloves are still known in the trade.

" Twelve pare of gloves " is an entry in Sir Thomas Boynton's inventory, 1581, and in some extracts from the *Household Book of Lord North* (*Archæologia*, vol. 19) there occur the following entries :

1578—For a peticote, vjli. ; gloves, xijs. ; for buskins, xxxxs. ...viijli. xijs.
1579—A riding clocke [cloak], lijs. ; doblets, ls. ; silk nether
    stocks [stockings], xls. ; for yarne hose, xxxs. ; ij
    hatts, xls. ; ij pair boot hose, xxiiijs. ; for camericke
    [cambric], an ell, xijs. ; for gloves, xxs. ; garters, vjs. ;
    sweet baggs, xxijs. viijd. ; for points, four dozen,
    viijs. ....xvli. viijs. viijd.

Another interesting series of old accounts, the *Household Book and Privy Purse Expenses of the Lestranges of Hunstanton* (*Archæologia*, vol. 25), contains several items of gloves.

1519.—Itm for a payer of hedgying gloves for ye carter ....iiijd.
1520.—Itm pd for vj payer of gloves for my master.....................ijs. viijd.
    Itm pd for ye makyng of Ms Besse gloves ....vjd.
    Itm pd for a payer of gloves bought at ye same feyer (Ely) ...jd.
1522.—Itm pd for a payer of glovys for my master ....iiijd.

and another pair at 3d. Still more interesting is a bill of

LADY'S EMBROIDERED GLOVES, SIXTEENTH CENTURY.
(In the possession of Rev. Walter Sneyd.)

the "expenses of the two brothers, Mr. Henry and Mr. Wm. Cavendish, sons of Sir William Cavendish, of Chatsworth, Kt., at Eton College, commencing October 21, 1560 (*Retrospective Review*). The details are comprehensive enough, covering more than one payment for "mending their showes," and one sum of 6d. for their "quarterydge in penne, ynke, byrche and brome." Two pairs of "furred gloves with strynges at them" were charged fivepence.

A *Lesson of Wysedome for all Maner Chyldryn*, signed by one Symon (*MS. Bodl.*), full of injunctions for the little ones to mind their manners and not get into the mischief to which it would seem little mediæval boys were as prone as any other, enjoins that care shall be taken of their gloves—

> Chyld, kepe thy boke, cappe, and glouys—

and promises them, in case the caution is disregarded, a birching in the manner sanctioned by long usage, familiar, indeed, with some of us in the days of our youth, although generally disregarded now, unless the bircher would run the risk of a charge of assault.

In the *Bill of the Expenses attending the journey of Peter Martyr and Bernardinus Ochil from Basil (Basle) to England, in* 1547 (*Archæologia*, vol. 24), "for 2 payer of glovys for them" the sum of one shilling is paid, and for "a payer furryd gloves for P. Marter" 13s., as well as £1 11s. 3d. "for a peticote, glovys, and nyght cap for Julius," the servant of Peter Martyr. Gloves were now common to all classes and conditions of men. "Slefes, purses, and glofs" are included in an inventory of the effects of Anne Nycolson, of "Kyrby," who died in 1557. Forty years later, in-

LADY'S EMBROIDERED GLOVES, SIXTEENTH CENTURY.
(In the possession of Rev. Walter Sneyd.)

cluded in the extensive stock of John Farbeck, a wealthy
Durham mercer, are found—

xxx whit gloves, viijs.
vj pare of fyne gloves, xjs.
(SURTEES SOC., *Wills and Inventories.*)

and we find Mistress Joyce Jefferies yet later buying
" Cordovan " gloves, sweet gloves, and gold embroidered
gloves (*Arch.*, vol. 37).

The gloves of the sixteenth century are well illustrated
in the accompanying engravings. The magnificent embroidery on the cuff of the glove, of which both back and
front are given, can hardly be done justice to in description,
or even in a colourless print. Every flower, the columbine
and the pink in particular, the butterflies, and even a little
goldfinch in the middle of the cuff, are rendered in natural
colours with an exquisite fidelity, and with such skill as to
make them veritable needle-paintings, in which, too, the
needle well holds its own against the brush. The work is
done in fine silk, and the shading is eloquent of the skill of
early dyers, for the range of colours admitting of such undefinable gradations must have been very extensive. The
colours are, of course, somewhat faded, but, considering
their age, are wonderfully well preserved. The raised gold-
work and stitching with gold thread are also in excellent
condition, though the work has in some places worn out the
white satin on which, with such excellent skill, it was first
grounded. The glove is nearly thirteen inches in total
length. The whole cuff, four-and-a-half inches in depth, is
lined with crimson silk, and the side bands of cloth-of-gold
ribbon, edged with gold fringe, were probably attached to
the gloves to confine the wide sleeves, and allow the orna-
mentation of the gauntlets unhindered admiration.

The other gloves (*see next page*) are very notable ones. Possibly they represent those more in ordinary wear, and

SHAKESPEARE'S GLOVES.
(In the possession of Miss Frances Benson.)

may be taken as fair specimens of common sixteenth-century gloves, but they have extraordinary interest, for they are believed to have belonged to Shakespeare. It is natural to regard such relics with suspicion ; and the higher the claim to veneration the stronger is the doubt which must first be overcome. There is in this case, however, good presumptive proof that these are worthy of such tender respect as we must feel for any personal association with the immortal poet. Accompanying them there is a memorandum as follows :—

A Pair of Gloves worn by Shakespeare.

Presented to Garrick by the Mayor and Corporation of Stratford-on-Avon, at the time of the Jubilee there (1769), in a finely carved box of the Mulberry Tree planted by Shakespeare; together with the lease of his house in Stratford, and the freedom of the town.

Garrick's widow presented these to Thomas Keate, Esq., of Chelsea College, about 1800, and his daughter in turn gave them to her relative, the mother of Miss Frances Benson, in whose possession they now are, although the box is unfortunately lost. There is an unimpeachable testimony to the identity of the gloves from the hands of Garrick downwards ; and if the evidence on which the Corporation of Stratford believed them to have been worn by Shakespeare was equally trustworthy, there is no doubt that these are the very gloves they are represented to be.

These were real workaday gloves, and have plainly seen some wear. Made of substantial leather, they are not altogether destitute of ornament. The scroll stitching on the knuckles has been in red and gold, two colours maintained throughout all the accessories of the glove. The ribbon marking the cuff is of yellow silk, and that on the lower edge of crimson, with a yellow fringe. The cuff is of double leather, with a pattern pinked in the upper skin.

Gloves of a more elaborate order of the same period

MR. CARTAR'S .GLOVE.

(In the possession of Miss Mary Mayo.)

GLOVE (SEVENTEENTH CENTURY.)
(In the possession of Miss Mary Mayo.)

appear in the accompanying illustrations. Both are of tan-coloured doeskin lined with white kid, and in the cuff with red silk turned up over the edge. One is inscribed within :—

Mr. Cartar's Glove,
Yarmouth.
His picture was taken
1627.

but Mr. Cartar's glove has outlived his memory.

The other glove is embossed with gilt twist and gold thread sewn down, on red satin appliquè, with the conventionalized heartsease in the natural colours of the small wild flower. We may fairly assume that, gorgeous as these gloves are in comparison with those now generally used, they were not uncommon in the time from which they date, or we should not find plain Mr. Cartar wearing them.

Pedlars now included them among their wares. These, the welcome gossips at many a lonely grange and manor house, were then honoured members of an esteemed profession, transacting considerable business at leisure and in comfort. This made Wordsworth—who, by his *Excursion*, was clearly favourable to the life—often say that he would, had it been necessary, have gladly chosen to be a pedlar, as the calling was once carried on. In the seventeenth century the pedlar stood high in popular esteem, at the summit of his social condition. Railways have ruined his calling, and the ways of the " credit draper " spoilt his reputation. In the beginning of the seventeenth century pedlar's songs were in great favour, and obviously afforded opportunities of giving an alluring list of goods and finery. One of these is included in John Dowland's

Seconde Booke of Songs or Ayres, 1600, and the presumptive pedlar sings—

> Within this packe are pinnes, pointes, laces, and gloves,
> And divers toies fitting a country faier.

And another from *Catch that Catch Can*, 1652, says—

> Come, pretty maydens, what is't you buy ?
> See what is't you lack ?
> If you can find a toy to your mind,
> Be so kind, view the pedlar's pack :
> Here be laces, and masks for your faces,
> Corall, jet, and amber,
> Gloves made of thread, and toys for your head,
> And rich perfumes for a lady's chamber.
> Come and buy ; come, buy for your loving hony,
> Some pretty toy,
> To please the boy,
> I'll sell you worth your money.

Both of these are quoted in Fairholt's collection previously acknowledged, as well as a thirteenth-century song of a French mercer, which announces :—

> J'ai beax ganz à damoiseletes
> J'ai ganz forrez doubles et sangles.

Another black-letter ballad, printed for *I. Back*, at the Black Boy, on London Bridge, is entitled " The Sorrowful Lamentation of the Pedlars and Petty Chapmen for the Hardness of the Times and the Decay of Trade," and has the following verse :—

> We travail all day through Dirt and through Mire,
> To fetch you fine laces and what you desire :
> No pains we do spare, to bring you Choice Ware,
> As Gloves and Perfumes, and sweet Powder for hair.
> *Then, Maidens and Men, Come, see what you lack.*
> *And buy the fine toys that I have in my Pack.*

During the Stuart period gloves of great length were worn. The prints of Hollar and Silvester afford ample proof of this. As sleeves receded, gloves advanced, and when the sleeves had become mere shoulder-puffs, gloves had gone beyond the elbow after them. Dress is usually a matter of development, and it almost invariably happens that these two parts of it are in strict relationship to each other. The gloves of the gay cavaliers were generally of white leather with wide cuffs, but they were often over-loaded with ornament, illustrating another law of fashion which makes one extreme follow another as surely as the strokes of a pendulum. From the plain straight dress of the grim Puritan, the Restoration brought out a flowing voluminous garb with all manner of pendants and plenty of room everywhere, from the wide-brimmed hat and nodding plumes to the tunnel-topped and lace-trimmed boots. Ribbons were used in wonderful profusion, till Evelyn says of a fop, "a fine silken thing" that he saw in Westminster Hall, that he had "as much ribbon about him as would have plundered six shops and set up twenty country pedlars. All his body was drest like a May-pole or Tom-o'-Bedlam's cap." Gloves had these adornments. Ralph Thoresby had, in his Museum, three or four sorts of these gloves "top'd with narrow Ribbands of various colours and textures with gold or silver interwoven. . . . White gloves with broad black lace ruffles and heavy fringe ; gloves pearl-colour and gold ; these were used in my own time." Women's of the same time (ult. Chas. II.) "had large rolls of ribband round the tops and down to the hand, plain crimson satten, intermix'd with strip'd and flowered, edged with gold " (A. Th.'s *Wedding Gloves*). Thoresby had, too, in his Museum, a pair " of fine Holland with black Silk Needlework and a wrought lace of both colours," which remind us of those "linning stockings" with which good master Pepys made himself "as fine as he could." Thoresby says, too, that gloves such as those worn by James I. were in the next reign worn by private gentlemen :—

Witness a pair of my wife's grandfather's, richly embroidered upon black silk and a deeper gold fringe. The embroidery reaches above the elbow. Another pair somewhat shorter, embroidered upon the leather, lined with crimson silk. They were Mr. *Fran. Layton's*, who was of the Jewel House to *K. Charles I.*, the gift of his son, *Tho. Layton*, Esqre. A pair of the common size, but richly embroidered, with raised or emboss'd work, when Mr. *Geo. Thoresby* was Sheriff of Newcastle, in Northumberland. His wife's, which are deeply escaloped, have black Bugles intermixed.

In 1624 Sir John Francklyn paid 10s. for "two pare of thick gloves," and 7s. for "two pare of thin." The *Mercurius Publicus* of July 5, 1660, contains an advertisement setting forth the contents of a "Leathern Portmantle lost at Sittingburn or Rochester when his Majesty came thither,

LADY'S GLOVE, TEMP. JAMES I.
(ASHMOLEAN MUSEUM.)

K

wherein was a Suit of Camelot Holland, with two little
laces in a Seam, light pair of white gloves, and a pair of
Doe's leather." Lace was freely used on gloves, a fashion
which appears at this time to have been started in France,
although they had been so ornamented more than a
century before, for Mary Basset, daughter of Lady Lisle,
in 1538 writes to her mother gratefully acknowledging
" the laced gloves you sent me by bearer." But in 1661,
Madlle. de la Vallière attracted attention at a fête at Vaux
by wearing gloves of cream-coloured Brussels lace ; and
here, in spite of prohibited import, gloves as well as all
manner of minor articles, were lavishly trimmed with costly
laces. In this, royalty took the lead. Charles II., in 1661
by proclamation re-enforced an Act of his father forbidding
foreign lace to be brought into the country, but in the
same year granted licence to John Eaton, allowing him to
import such quantities of lace made beyond the seas " as
may be for the wear of the Queen, our dear mother, the
Queen, our dear brother James, Duke of York, and the rest
of the royal family, to the end the same may be patterns
for the manufacture of those commodities here, notwithstanding the late statute forbidding their importation."
This attempt at exclusiveness defeated itself, by giving
fresh zest and irresistible impulse to efforts made to obtain
by foul smuggling what could not be procured by fair and
open trading.

At this time the Marquis of Worcester, in his *Century of
Inventions*, which so remarkably anticipates many modern
scientific achievements, would have endowed gloves with a
language so as to make them available for secret correspondence. Invention 34 is " A glove with knotted silk
strings ;" 35 " A glove with fringes and knotted silk

strings;" and 37 "A glove pinked with an alphabet."
"The knots shall signify any letter with commas, full
points, and interrogations, as legible as with pen and ink
and white paper." The ingenious nobleman apparently
knew well the value of forming varying combinations of
arbitrary symbols, not only for cypher correspondence, but
in less reputable employment; for Invention 89 is for
"White Silk knotted in the fingers of a Pair of White
Gloves, and so contrived without suspicion that playing at
Primero at Cards one may without clogging the memory
keep reckoning of all Sixes, Sevens, and Aces' which he

FROM PORTRAIT OF THOMAS CECIL,
FIRST EARL OF EXETER,
BY JANSEN, 1621.

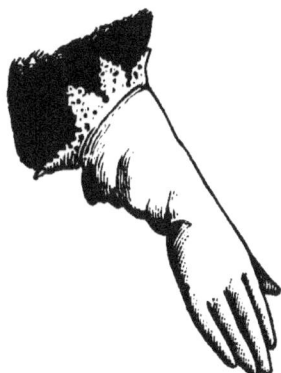

FROM PORTRAIT OF HENRY WRIOTHESLEY,
EARL OF SOUTHAMPTON,
BY MIREVELT 1624.

hath discarded." This certainly has something of the
flavour of cheating, but is, after all, only substituting an
infallible mechanical memory for one probably faulty.
Whether this would be taking an unfair advantage of an
opponent is a question of no moment; but as regards the
honour of the noble Marquis we can hardly hold him in
great respect when his next invention is for "A most dexterous Dicing box;" so that, whatever his practice may have
been, his principles can hardly be considered unimpeachable.

The fringed gloves which came in during the reign of Charles II. were very fashionable all through the first half of the eighteenth century. Addison professed to have received "a heavy complaint against fringed gloves," and No. 311 of the *Spectator* mentions "a young jackanapes with a pair of silver-fringed gloves" as being caught paying attention to a handsome heiress. With a common inconsistency, these bullion fringes were worn on gloves of ordinary leather. They were worn, too, on the coats and waistcoats of beaux when George the Third was king.

Beyond this point there is nothing to chronicle in the history of gloves, which have gradually become more and more severe in their simplicity, knowing little change except in their length, and having their chief merit in an immaculate fit. This is so far recognized as the first essential of a glove that some superfine exquisites have had their gloves made on prepared models of their hands— really hand-lasts. Another trifling innovation in gloves is the introduction of a purse glove, so that ladies should not have to fear danger either from purse-snatching thieves, or their own negligence when, rapt in the delights of shopping, the *porte-monnaie* is left behind on some unfriendly counter. This is in some respects no new thing, and, like many another brand-new novelty, only provokes recollection of remote customs. For instance, Du Cange mentions *manufollia*, mittens filled with money and laid under the pillow ; and, according to Fosbrooke (*Cyclopædia of Antiquities*) it was once the custom to keep money in gloves, so that the long stocking has usurped somewhat of its widespread reputation as a homely bank.

M. Challamel records with characteristic vivacity how it

was proposed, a few years ago, to dispense with gloves altogether :—

A strange rumour was current in the highest circles in 1873.

What was that ?

Nothing less than the abolition of gloves ! This was assuredly no question of economy, for their place was to be taken by a fashion worthy of the days of the Directory. Women of fashion proposed to wear clusters of rings between every finger-joint ; each hand to bear a fortune.

This was the fantastic dream of some blasée fine lady, longing for novelty at any price. It was not realized, as may be imagined ; and gloves kept their place—an important one—among articles of feminine attire.

This dream has been realized in America. For several seasons, it is said, New York men refused to wear gloves at balls, on the score that the Prince of Wales had discarded them. All the sarcasm lavished on the republican love of titles and the reverence in which a lord is held by "citizens" never equalled this. The persuasions of their fair partners availed nothing against the example of the heir to a throne. There may, in some cases, be a principle involved in appearing in ungloved hands. George Stephenson, on being urged to wear gloves when about to be admitted to an audience with the King of Belgium, said that he was only a plain man, and if the King of the Belgians could not receive him in nature's gloves, cleanwashed, he would not go at all. Gloves, on such occasions, may be neglected for want of knowing better; as with a navy captain at a Portsmouth ball, who, when his partner, a lady of rank, suggested the propriety of putting on his gloves before they led off, remarked, " Oh, never mind me, madam! I shall wash my hands when I have done dancing." (*Quarterly Review.**)

* Lest this reference should be regarded as an error or a hoax, the reader may be directed to the *Quarterly Review*, vol. lix. p. 410. It would, however, be doing the reviewer an injustice not to mention that a volume entitled *Hints*

on Etiquette is there under consideration, and more particularly an injunction stated to be given on the authority of an anonymous "lady of rank," who allowed the author free access to her note-book.   Her ladyship's instructions run thus, the very italics being her own :—

"Do not insist upon pulling off your glove on a very hot day when you shake hands with a lady.   If it *be off*, why, all very well ; but it is better to run the risk of being considered ungallant than to present a *clammy*, ungloved hand."

Against this opinion, on a point the gravity of which must be patent to everybody, the reviewer quotes a passage from the work of one James Pitt, professor of dancing and author of *Instructions in Etiquette :*—

Q.—Is it proper, on entering a room, to take off the gloves to shake hands with the company ?

A.—It will always be correct for gentlemen to take off the glove of the right hand ; but ladies are allowed to keep on their gloves ; nevertheless I should not advise them to avail themselves of their privilege when they wish to show respect, and especially to an intimate friend ; *for friendship is so sacred, that not even the substance of a glove should interpose between the hands of those who are united by its influence.*   Be careful, in taking off the glove, that you do so with ease and grace, avoiding all appearance of attending to your hand when you ought to be attending to your friend.

CHAPTER XI.

Companies of Glovers.

WHAT time glovers banded themselves together for their mutual protection and profit cannot be fixed with certainty. Probably as soon as gloves became commonly worn, and the trade of making them a lucrative calling, there would be an association of its members. Craft gilds, societies of handicraftsmen, came into existence in the twelfth century, when English trade was gaining weight, but it was not until the fourteenth century that these obtained the mastery, after a bitter struggle, over the older burgher gilds which they superseded. With us, glove-making was first carried on by the tawyers, skin-dressers (Anglo-Saxon, *tawian*, to prepare), and the trade appears to have had no distinct boundary line from cobbling. Tawyers, cobblers, and cordwainers—workers in leather made in imitation of that of Cordova in Spain—all appear in one division in the *Liber Albus*. In the fourteenth century it was ordered that "a pair of shoes of cordwain shall be sold for sixpence ; one pair of cowhide for five pence ; one pair of boots of cordwain for 3s. 6d. ; one pair of cowhide for three shillings ;" and "one pair of gloves of sheep leather for one penny halfpenny, and the best for two pence."

In France the glovers had formed themselves into a company at least as early as 1190, when they were under the rule of a settled code of statutes. These laws received confirmation on various occasions by the kings of France, and were renewed, confirmed, and enlarged by Louis XIV. under letters patent in March, 1656 (SAVARY: *Dictionnaire Universal de Commerce*). The patron saint of the French glovers is St. Anne, presumptive mother of the Virgin Mary. According to tradition, St. Anne was a knitter of gloves, winning subsistence by following that occupation, and her memory is held in tender veneration by all good glovers in France. At any rate, until a recent period her festal day was observed with great solemnity in all the principal seats of the French glove trade, and particularly at Grenoble. The officers of the London Company of Glovers were always formerly elected annually on the 8th of September, the feast of the Nativity of the Virgin Mary, so that the tradition was plainly also held here in loving remembrance.

The Incorporation of Glovers of Perth are declared to have received from William the Lyon (1165) a charter " admitting them to the free right and privileges of being merchant burgesses, in addition to their own peculiar rights as incorporate craftsmen." When Edward I., after his conquest of Scotland, directed that all the archives of the country should be overhauled, and all documents of national import be sent to England, this charter, together with the original deed of incorporation which it is believed to have confirmed, were included among the spoil, and went down, lost for ever, in the vessel that was carrying the greater part of the documents to London. The Scotch glovers have, supposing these deeds to have been in existence and lost on this occasion, genuine cause for regret that the high antiquity of their trade cannot be conclusively established.

The glovers of the ancient Scotch capital had for their patron, St. Bartholomew. A former deacon of their Incorporation, Andrew Buist, contributed to Penny's *Traditions of Perth* a paper of considerable value relating to the affairs and history of the "calling" of which he was obviously proud to be the head. The connection of St. Bartholomew with glovers he plainly ascribes to the legendary accounts which show the saint to have been flayed before his crucifixion. To the saint the calling erected an altar in the church of St. John the Baptist, one, and that among the richest, of several shrines in this parish church of Perth ; several others of which were established and maintained by the incorporated trades of the city —the Weavers dedicating one to St. Severus, the Hammermen another to St. Eligius or Eloi, the Fleshers (Butchers) to St. Peter, and the Cordiners (Cordwainers) to SS. Duchàne, Crispin, and Crispinian. The Glovers of Perth (who are divided into two "sciences," Skinners and Glovers) still show a picture of their patron saint, dated 1557, which, after long neglect, has been recently restored and hung up in their hall. This shows the saint in a sitting posture, with the open Bible before him and a flaying knife in his hand, and also displays the tools of the craft, knife, shears, and bodkin. Another accessory of the picture is a view of a house—a "mean-looking tenement" in the Curfew Row, or "Couvrefew Street," as Scott wrote it—still known and visited by pilgrims to the shrines of fiction as "The Fair Maid's House." This was the old Glover's Hall, and according to popular tradition was once a chapel dedicated to the saint—a view which Mr. Fittis, who recently contributed a series of articles on the subject to the *Perthshire Advertiser,* declines to receive. A niche in the house in which, according to Mr. Morrison—the "local antiquary to whom Scott acknowledges his deep obligation"—was once hung the curfew bell is believed, too, rather to have held some image or emblem of the craft whose meetings were held there until 1787, when a new hall was built for the purpose. Whatever may be the facts as to these matters in dispute, they all go to prove how Scott has peopled Perth with characters that have yet an abiding interest, and endowed the city and the glover calling with memories which will never fade.

Several of the articles of historic interest noticed by the great novelist, either incidentally or in the notes attached to the novel, yet remain. There is St. Bartholomew's Tawse—"the whip of St. Bartholomew, which the craft are often admonished to apply to the backs of refractory apprentices," and not to them only, but to hired workmen as well. This instrument appears to have been in frequent requisition. There is, too, the flag under which the crafts-men—who were as ready with the spear as the needle—often did true yeoman's service for "bonnie St. Johnstone." It was quite fitting to make old Simon Glover boast that, when war came to the gates of the fair burgh, "down went needles, thread, and shamoy leather, and out came the good headpiece and target from the dark nook and the long lance from above the chimney." Deacon Buist shows the flag to have in his time borne an incomplete motto, the gold letters in which it was inscribed, or the substance which attached the letters, having corroded the silk and made it drop. The motto as it was restored some fifty years ago reads—

The Perfect Honour of a Craft or Beauty of a Trade is not in [Wealth but in Moral Worth] whereby Virtue gains renown :

the brackets marking the restoration of or substitution for the words which Deacon Buist laments as lost. The motto, in an ellipse, encloses the arms of the Incorporation, a pair of gloves "displayed" on a shield, with three green-coloured stars surmounting them, and the motto "Grace and Peace" in a scroll over all.

The Glovers still show, also, the Morice-dancing dress noticed by Scott, a curious garment composed of fawn-coloured silk, and made in the form of a tunic, with slashed sleeves, and trappings of red and green satin. There accompany it two hundred and fifty-two small circular bells, arranged in twenty-one sets of twelve bells each, upon pieces of leather, intended to be fastened on various parts of the wearer's body. What is most remarkable about these bells is the perfect intonation of each set, and the regular musical intervals between the tones of each cluster. The twelve bells on each piece of leather are of various sizes, yet all combining to form one perfect intonation in accord with the leading note in the set. These concords are maintained, not only in each set, but also in the intervals between the various pieces, so that the performer could thus produce, if not a tune, at least a pleasing and musical chime, according as he regulated with skill the movements of his body. In this Morice-dance the Glovers were particularly proficient ; it was with them "our dance," and they were at times, when royal personages visited the city, called upon to exhibit their skill for the diversion of the distinguished guests. James VI. was so entertained in 1617, when the city magistrates paid the Company £40 to recompense them for expenses proper to the occasion. Charles I. visited Perth sixteen years after, and a "Memorandum" in the Glover records relates, artlessly enough, the proceedings, in which they again danced before the King.

At the entry of our South Inch Port, he was received honourably by the Provost, Bailies, and Aldermen, and by delivery of a speech mounting to his praise, and thanksgiving for his Majesty's coming to visit this our city, who stayed upon horseback, and heard the same patiently; and therefrom convoyed by our young men in guard, with partisans, clad in red and white, to his lodging at the end of the Southgate (Gowrie House), belonging now heritably to George, Earl of Kinnoull, High Chancellor of Scotland. The morrow thereafter (the King) came to our church, and in his royal seat heard ane reverend sermon. Immediately thereafter came to his lodging, and went down to the garden thereof. His Majesty being there set upon the wall next the water of Tay, whereupon was ane floating stage of timber clad about with birks, upon the whilk, for his Majesty's welcome and entry, thirteen of our brethren of this our calling of Glovers, with green caps, silver strings, red ribbons, white shoes, and bells about their legs, shewing rapiers in their hands, and all other abulziement, danced our sword dance, with many difficult knots and allafallajessa, five being under and five above upon their shoulders, three of them dancing through their feet and about them, drinking wine and breaking glasses. Whilk (God be praised) was acted and done without hurt or skaith to any. Whilk drew us to great charges and expenses, amounting to the sum of 350 merks, yet not to be remembered, because graciously accepted by our sovereign and both estates, to our honour and great commendation.

Afterwards, Mr. Fittis says, this dress was, on a regiment being raised by the Duke of Athole, during the War of Independence, brought out in the train of the Deacon of the Glovers as he went about the town recruiting to make up the quota which each trade had promised towards the "Athole Highlanders;" and again, when Lord Lynedoch was raising, in 1794, the 90th Regiment of Foot—the "Grey Breeks"—he did not disdain to stimulate ancient memories by wearing this quaint dress through the streets of Perth when on recruiting bent. The Morice costume was again donned by a member of the craft, who, with a company of his brethren, appeared before the Queen and the Prince Consort on their first visit to Perth in 1842.

There is no anachronism in making Simon a Glover, both by name and occupation, in the days of Robert the Third, of economical memory, under whose rule occurred the events on which *The Fair Maid of Perth* is founded. Simon Glover is no mythical or fictitious personage. At the submission of the Magistrates of Perth to the first Edward, in 1296, among the company of city notabilities who came forward to take the oath of fealty were John de Perth, the provost, Bernard le Mercer, and Simon le Glovere (PALGRAVE: *History and Affairs of Scotland*). Simon of the novel boasts himself the son of a glover, and is well stricken in years, so that he had entered upon a period when, as students of our nomenclature know, surnames derived from trades or residences, from nicknames or personal peculiarities, were sticking to people like burrs to cloth, and marking more clearly the boundary-lines of relationship. Not long afterwards we find glovers following other occupations, as well as men of foreign surnames becoming glovers ; like the John Wylkynson of Hadley, Suffolk, who in 1488 received the king's pardon (*Record Papers*), or John Vyntcent, glover of Lynn, who, in 1519, paid the L'Estranges, of Hunstanton, seven shillings for "xxviij shepe skynnys," and in the following year 9s. 2d. for 33 more skins. Another glover of Walsingham is also shown as a purchaser of sheep skins. (*Arch.* xxv.)

One John le Gaunter, probably a glover, is recorded to have slain Alexander de Holebeck, in 1242, in a brawl ;

and this is, in all probability, the first record we have of any English glovers. The first mention of a Company of Glovers in England occurs among the performers of *A Play of the Old and New Testament*, exhibited at Chester, in 1327, at the expense of the different trading companies of that city, then a place of considerable commerce. This . was one of the Miracle Plays, at which Chaucer's Wife of Bath attended during Lent. These plays often grossly belied their title of Moralities; the fact that they represented Scripture scenes, often played by monks, not preventing, but rather often provoking some dreadful enormities. On this occasion, the various scenes were apportioned among the several companies in the following order, as appears by a MS. in the Harleian Collection (2013)—*The Fall of Lucifer*, by the Tanners. *The Creation*, by the Drapers. *The Deluge*, by the Dyers. *Abraham, Melchisedek, and Lot*, by the Barbers. *Moses, Balak, and Balaam*, by the Cappers. *The Salutation and Nativity*, by the Wrightes. *The Shepherds Feeding their Flocks by Night*, by the Painters and Glaziers. *The Three Kings*, by the Vintners. *The Oblation of the Three Kings*, by the Mercers. *The Killing of the Innocents*, by the Goldsmiths. *The Purification*, by the Blacksmiths. *The Temptation*, by the Butchers. *The Last Supper*, by the Bakers. *The Blind Men and Lazarus*, by the Glovers. (WARTON : *Hist. English Poetry*).

In a procession of trade pageants, which was once annually carried out on Corpus Christi day, the " Glovers and Skinners or Guild of St. Mary " are placed thirteenth in order of precedence. On one occasion it was appointed to them to represent " Adam and Eve with an angel bearing a sword before them," when they would be enabled to make a more endurable appearance than at

Chester, where the order of the play directs our first parents in due sequence to "appear naked and not ashamed." It is stated that these instructions were usually carried out to the letter, but it has been proved that the performers did not appear in such primitive fashion, but were clothed in a tight-fitting garment of white leather. (HALLIWELL-PHILLIPS : *Outlines of the Life of Shakespeare.*) The accounts of the expenses on these occasions, from the characters assumed by the players, contain some items so strange as often to shock reverential feeling. Passing these by, we find in one of the *Coventry Mysteries*, likewise performed on Corpus Christi day, a payment of two shillings "to the Marie for her gloves and wages ;" and a writer in the *Transactions of the Royal Historical Society* quotes another bill of strange expenses, incurred in the fourteenth century, on another of these displays at Coventry. Among these are "Payd for two pound of hayre for the divell's heed," 3s. ; "mending his nose," 8d. ; "red buckram for wings of angels," 7s. ; and "Payd for a cote and payr of gloves," 3s. In these representations Herod, like the Doges of Venice when they figuratively wedded the Adriatic, wore generally red gloves. The accounts of the Smiths' Company, in 1502, show, "paid for gloves to the pleyares," 19d.; and the Cappers' Company, in 1567, "payd to Pilat for his gloves," 2d.; and in the following year, again, "paid for Pylatt gloves," 4d.

The Glovers came out and were separate in the middle of the fourteenth century. The goods were plainly in demand, and the occupation big with business, so that they could assume airs of importance and ask to be recognized as a craft of some weight. We can fix the actual date of their assumption of this dignity by the "Articles," still preserved among the City archives, which defined the position of the mystery and raised it above the rank of an ordinary trade guild, such as it would until this time have been. These "Articles" are of deep interest. Not only do we read, in them, the regulations and restrictions of this particular craft, but we read, beneath them, the prevailing spirit of trading legislation at those times, and follow the strictly-defined and carefully-guarded channel in which trade was made to flow.

These are the points and ordinances which the good people, Glovers of London, demand to have and hold firm and established for ever to the saving of their mistery and to the great profit of all the common people : —

Imprimis, that no foreigner of this mistery shall keep shop or use this mistery, or sell or buy, if he be not a freeman of the City.

Item, that no one of this mistery shall be received into the franchise of the City without the assent of the wardens of the same mistery or of the greater part of them.   Item, that no one of this mistery shall take or entice a servant of another from the service of his master so long as he is bound by covenant to serve him, on pain of [paying] 20 shillings to the use of the Chamber, if before the Mayor and Aldermen he be convicted thereof by the men of the said mistery.

Item, if any servant of the said mistery make away with the goods or chattels of his master to the value 12 pence more or less, such default shall be redressed by the good men the wardens of the said mistery ; and if that servant who shall have so offended against his master shall be unwilling to submit to be judged by the wardens of the said mistery, let him be forthwith attached and brought before the Mayor and Aldermen, and before them let the default be punished according to their discretion.

Item, that no one of the said mistery shall sell by night by candle-light in any house their wares, since people cannot have such good cognisance by candle-light as by light of day whether the wares be made of good leather or of bad, or well and lawfully, or falsely, made upon forfeiture of the wares sold by candle-light to the use of the Chamber.

Item, if any false work touching the said mistery be found or brought for sale within the Franchise of the said City, let it be forthwith taken by the wardens of the said mistery and brought before the Mayor and Aldermen, and before them this work adjudged to be such as it shall be found to be, upon oath of men of the said mistery.

Item, that all things touching the said mistery sold between   oreigner [non-freeman] and foreigner, shall be forfeited, according to the ancient usages of the City.

Item, that every servant of the said mistery who works by the day shall not take more for his labour and work in the said mistery than he was wont to take two or three years before these points and ordinances were accepted by Walter Turke, Mayor, and the Aldermen—that is to say, the Monday next after the feast of the "Tiphanie" [= Epiphany, 6 Jan.] the 23rd year of the reign of King Edward the Third after the conquest. [1349—50.]

Item, whereas some persons who are not of the said mistery do take and appropriate to themselves the servants of men of the same mistery and make them work in secret in their houses and make gloves of rotten (?) and bad leather, and do sell them wholesale to strange merchants coming to the City, in deceit of the people and to the great scandal of the good men of the said trade, that the wardens of the mistery make search in such manner for gloves made of false material that the same may be found, and carry them before the Mayor and Aldermen ; and before them let them be adjudged to be such as they shall be found to be, upon oath of the good men of the mistery.

Item if anyone of the said mistery be found obstinate to act contrary to the points aforesaid or any one point of them, let him be attached by a serjeant of the Chamber at the suit of the wardens of the said mistery, to appear before the Mayor and Aldermen; and before them let him be punished at their discretion.

And be it known that all the under-written were elected by the wiser and wealthier men of the mistery aforesaid to safe-guard the articles aforesaid touching the mistery aforesaid, and were sworn before the Mayor and Aldermen to safe-guard the said articles, to wit

ROBERT DE GOLDESBURGH	
THOMAS DE GLOUCESTRE	
JOHN DE NORWYCHE	Sworn for the safe-guarding
JOHN LE BARBER	the articles above written.
WILLIAM DE DERBY	
JOHN DE WODHULL	

The idea of Free Trade was plainly not entertained— perhaps not even conceived—in the days of which this document stands so plainly representative. Nothing is so conspicuous in early English commercial history as the jealousy of "foreigners," which is so curtly and forcibly urged on behalf of native glovers. It is noticeable that

L

the disabilities of the foreigners, to whom English trade owes so much, were even in the middle of the fourteenth century sanctioned by " ancient " usage, and stipulations to this end were re-enacted again and again with renewed force and restrictions more and more harassing, and the business by which we, no less than the foreigner, received benefit was crippled by every possible device and sometimes forbidden altogether. At the time of these " Articles " no foreign merchant was allowed to reside in England without special license from the king, and, although this provision was relaxed almost immediately afterwards, the spirit of trade was, for some centuries subsequently, to shackle all aliens with liabilities and prohibitions from which the native-born merchant was exempted. It was not until the reign of Charles II. that the foreigner was allowed fair play in his trading here.

The regulation of wages by enactment also continued long after this date, even within hail of the present century. In the year in which these Articles were drawn up the Mayor of London issued a regulation fixing the wages and prices in all trades in the City. The frauds which are noticed may be considered perennial, and the particular provision against selling gloves by candle-light marked a common evil which found vent in other devices. An Act of Richard III., for instance, forbade merchants to spread before their shops or booths red or black cloths or any other thing by which the sight of the buyers might be deceived in the choice of good commodities.

It appears to have been possible, in the infancy of trade, to entrust to each craft—when the term had not acquired its invidious meaning—the task of ensuring honesty of workmanship, and a very common privilege given to companies was the right of search for any inferior or unwarrantable wares.* The Grocers had an officer whose duty it was to see that all drugs and groceries were duly garbled (picked) and cleansed before being offered for sale. The Merchant Taylors made inquisition into the length of yard measures. Indeed, the punishments inflicted not infrequently for various frauds in merchandise prove that trade was well looked after and controlled, and the Companies had a keen sense of their duty, no less than of the advantage to be gained by insuring excellence of material or manufacture. Beyond this, a general supervision over the trade of the City was given to and demanded from the higher officials of the Corporation. The Aldermen were directed to view, each in his own ward, the seals which bakers were compelled to place on their loaves, and were made inspectors of the measures of "Taverners." The sheriffs, too—the "eyes of the mayor," as they were called—took oath that they would cause to be kept the assizes of bread and ale, which regulated the quality and price of those provisions. The Ale-conner, who may be regarded as the father of excisemen, took oath—probably without any scruple—that he would be ready to taste any ale of a brewer or brewster when required to do so. It is worthy of mention that this primitive method of testing malt liquors, known as "palating," continued in use until the time of the Stuarts.

Robert de Goldesburgh and his colleagues did, at any

---

* The privilege of search was also delegated to the Mayors, Bailiffs, Headboroughs, and Lords of Liberties of Boroughs and Corporate Towns, as well as —and particularly in the case of leather—to the Chancellors, Vice-Chancellors, Proctors, Taxors, and Scholars of the Universities of Oxford and Cambridge.

rate for a time, check frauds and abuses in their trade. The *Liber Albus*—the White Book of the City of London— compiled during the mayoralty of Whittington by Richard Carpenter, of which we have in these latter days been favoured with a translation, sets forth in one division a brief summary of records of punishments inflicted for various malpractices in trade. Among these, so various as to show that ingenuity in swindling and artistic adulteration are peculiar to no age or time, we find indicated another City record, containing an account of the burning of some " false " gloves, breeches, and pouches, when—

On Monday next after the feast of Saint Gregory the Pope [12 March], in the 24th year of the reign of King Edward the Third, etc., the men of the trade of Glovers who had been sworn to keep the articles of that trade, came and brought before Walter Turk, Mayor, and the Aldermen, 17 pairs of gloves found upon John Fraunceis, of Norhamptone. The said men of the trade of Glovers brought also 28 braels called " bregirdles " [belts or girdles], found upon divers men whose names are underwritten : namely, upon John de la Cusyn 2 braels, John atte Feile one, Thomas de Wayllyhs one, Richard le Pynnere 2, John de Astone 7, Richard de Salope 2, Alice Blake 3, William Tristram one, John Chapman 8, and John Ede one ; asserting that the said gloves and braels were of false fashion, and vamped up of false materials, in deceit of all the people, and to the scandal of the whole trade.

And examination being made of the said gloves and braels before the said Mayor and Aldermen, upon oath of the reputable men of the said trade it was found that all the gloves and braels aforesaid were false, and vamped up in a false fashion in deceit of the people and to the scandal of the trade. Therefore it was awarded that the said gloves and braels should be burnt in the high street of Chepe, near the Stone cross there ; and accordingly on the same Monday they were there burnt according to the award aforesaid.

On the same Monday also, at the suit of the men sworn of the Trade of Pouchmakers, by award of the said Mayor and Aldermen there were burnt 19 false pouches that had been found upon Peregrine de Lesschies, 12 false pouches found upon Haudekyn Stompcost, 19 false pouches found upon Peregrine Johansone, and 4 false pouches found upon Agnes de Salesburi.

The pouches were the purses which were in those days, when dress was destitute of pockets, attached to the girdle.

The Pursers were afterwards, in the reign of Henry VII., united with the Glovers, and the Ordinances of the combined company of Glovers, Pursers, and Leathersellers, with the petition which preceded their union, are still preserved among the City archives.

The " Ordinance " of the Glovers, another of the City records, is directed against Sunday trading—a misdemeanour which is, in point of antiquity, but in that alone, most respectable; for, while prosecutions on this account are not unknown in our own day, the laws of King Athelstan "forbade all merchandizing on the Lord's day " under very heavy penalties.

To the honorable men and wise, the Mayor and Aldermen of the City of London, supplicate the good folk of Glovers of the same City, that it may please you to grant them, to the honour of God, that no one of the said mystery, nor using that mystery, shall keep open shop or stall of the said mystery, to sell his wares in any part within the liberty of the said City on Sunday, or on any other great day of festival, on pain of paying to the Chamber of the Guildhall at the first default 3s. 4d., and at the second default 6s. 8d., and at the third default 10s., and at the fourth default that he hold neither shop nor stall open of the mystery at any time within a quarter of the year next after his fourth default, on pain of forfeiture of all his wares and imprisonment of his body, and that the defaults in the said mystery so found be presented by the wardens of the said mystery from time to time before the Mayor and Aldermen for the time being, and before them be adjudged.

Afterwards, on Saturday, the morrow of the Conversion of Saint Paul [25 Jan.], the 50th year of the reign of King Edward the Third [1375-6] the underwritten men were sworn and elected to regulate the mystery of glovers and for faithfully presenting defaults, namely :—

JOHN DERNEFORD } Glovers elected and sworn as
THOMAS HARE } aforesaid.
WALTER FULHARDY }

A law suit tried November 29, 1396, and shown in the *Borough Records of Nottingham,* lets in some light on the

working of the trade at this time, and so needs no excuse
for the quotation.

Thomas de Lenton, glover, makes a plaint of Thomas del Peek on a plea of
trespass and contempt against the Statute, that whereas the same Thomas del
Peek in the week next before the feast of the nativity of Saint John, in the
twentieth year of the reign of the present King, made an agreement here at
Nottingham with the aforesaid Thomas de Lenton, to cut and work 22 dozens of
the gloves of the aforesaid Thomas de Lenton, so that the same Thomas de
Lenton should have every week from the said Thomas del Peek 2 dozens of
gloves well and faithfully cut and worked until the Eve of Saint Martin then
next following, no week being wanting of the aforesaid 2 dozens in work so
that he should have all the aforesaid 22 dozens between the feasts of Saint
Martin and Michael, taking for the dozen 3d. until 5s. 6d. for leather bought
from the said Thomas de Lenton should have been paid back : the same
Thomas del Peek only cut 6 dozens and 4 pairs of gloves of the aforesaid 22
dozens and left 16 dozens and 8 pairs unworked,* which the same Thomas de
Lenton should have sold at the Fair of Lenton and of the working of the aforesaid Thomas del Peek, and he has never had them, but was deceived in default of
the aforesaid Thomas del Peek ; and so he says that the said Thomas has broken
the said agreement with him, whereby the same Thomas de Lenton is injured
and has received damages to the value of Twenty shillings, wherefore he enters
Suit, etc.

Of this case the conclusion is not given, although " the
said Thomas," who was defendant, appeared to make his
case good.

There follow here copies of several most interesting
trade regulations passed for their government by the crafts-
men Glovers of Perth.†   While these show in many respects
the jealous and exclusive spirit which appears in the gild

---

* The addition or subtraction here is somewhat faulty.

† For these transcripts I am indebted to the kindness of William MacLeish,
Esq., Clerk to the Glover Incorporation of Perth.   In acknowledging, very
heartily, this gentleman's courtesy and willing assistance in my quest, I only
make him representative of others from whom I have solicited illustrations or
information.   The kindness met with, almost invariably, under such circumstances, is the truest solace of study.   In only two out of numerous instances
have I met with refusals.

statutes already quoted, there is in the provisions which mark so strong a contrast with the times that be, a careful regard of good workmanship and a vigilant watch over the quality of materials, which is altogether praiseworthy. There is in these ancient instances so much that illustrates old trade customs and privileges that they cannot but be held valuable in evidence of the inherent interest of the history of commerce, even in its minutest details.

EXTRACTS FROM THE MINUTES OF THE GLOVER INCORPORATION OF PERTH.

Act anent prices for shewing Gloves.

16 *May*, 1598.—It is statute and ordained by the Deacon in a head Court holden at the Giltn Arbor,* in presence of the Haill [whole] Brethren of the Incorporation, that all their Servants from that date should have no more but the prices underwritten for shewing [sewing] of Gloves, and that no Master of the Craft should give any boy or other more than viz. :—For each dozn shewing shewed with silk, six shillings ; each dozen shiveringt shewed with threed, five shillings ; the doz. of Calf leather five shillings ; and the doz. of sheep leather at four shillings, certifying yt. the contraveeners of this order should pay fourty shillings unforgiven.

Act against going to Alehouses with work not being sent for.

12 *May*, 1615.—It is statute and ordained that if from the date hereof, it should be found that any freeman or freeman's servants go to a Tavern or Ale-house with Gloves or other work boding them upon Strangers, or others, they not being sent for, shall pay Five pounds Scots in to the Calling *toties quoties* [unforgiven.]

---

* Probably at this period a real arbour, or, as we may call it, summer-house, in which, as did other corporations of glovers, and notably those at Shrewsbury, the members of the craft would assemble first for some business, enough to qualify the occasion, and after for much more recreation. The Gilten Arbour at Perth was the name of a croft lying near the Monastery.

† *Qy.*—A corruption of *chevre*—goat-skin.

Against buying
any foreign work
or any work from
Landward
Skinners or
Chapmen
[Pedlars].

9 *November*, 1627.—Which day in a head Court. convened in the Craft's House anent the affairs of the Craft, they statute and ordain, that none of the brethren in time coming shall buy any Gloves or other work from any Landward Skinner or Chapman to sell forth of their Booths, or any English or foreign work, certifying that those who do in the contrary, and the same being tried and sufficiently verified, the goods shall be forfeit, and the party offending shall pay Five pounds of unlaw *toties quoties.*

Prices for
steeping Skins.

6 *March*, 1652.—Which day being convened in the Craft's House, the Deacon and bretheren of the Craft in a General Meeting, they ordain thirty two pennies Scotts money to be paid for every steeping of skins in any hole within the Bark House on the Leadside by the owners thereof to those who

Each freeman
to have his room
in the Barkhouse
as it falls to his
turn.

aught the Barkhouse, And Ordains every Brother of the Calling to have their room therein as it falls to their turn, and if the Master of the same let them to any Freeman or brother of the Craft the taxman shall have no more liberty than

No woolen skins
to be steeped
after 20th March
yearly.

any other Brother, but as it falls to his turn, As also Ordains that none of the Brethren steep any woolen skins in any of the holes on the Leadside after the 20th day of March yearly in time coming during the will and pleasure of the Calling, As

Prices for
shewing of
Gloves fixed.

Also Ordains fourteen shillings to be given for shewing the Doz. of out seam shiverings ; and twelve shillings for the Doz. of inseam Lamb Leather and inseam Sheep leather, and twenty four shillings for the Doz. of twice shewed sheep leather, and thirty two shillings for shewing of the Doz. of

Apprentices to
take work which
the Shewers cast.

Staig or Wild leather, and ordains every boy and fial to take such work from his Master as his shewers cast.

Unsufficient
work to be cut
and the Makers
thereof punished.

2 *May*, 1657.—Which day being convened in the Callings Meeting House, the Deacon and Auditors thereof representing the whole body and community of the said trade, Enacts and ordains, that none of the brethren in time coming shall make any insufficient work, certifying that such work shall be cut by the thumb, and the makers thereof punished otherwise as Deacon with the consent of the Auditors and Brethren as made in an Act heretofore.

NoGloves to be
sold before they
be searched.

27 *April*, 1663.—Which day by advice and consent of the whole Brethren, convened in a general court it is enacted that no brother of the Calling presume to sell or put away any made work of Gloves to any Merchant or Chapman furth of the Town at Markets until they be searched by the Deacon and Searchers nominate by the Calling yearly and that the Deacon

and Searchers be required by the sellers for that effect under the pain of Five pounds tó the Contravener *toties quoties.*

Whoever abuses work in the shaping or shewing thereof shall lose the same.

23 *May*, 1663.—Which day the Calling convened in their Meeting House in a general Court ordained and concluded that whatever Brother of the Calling who shall shape or shew insufficient work to any of the Calling, and taken by Searchers therefore shall be liable in the loss of the same and whoever shall abuse work in the shewing thereof shall incur the said penalty.

Act against shewing, colouring or whitening Gloves to Strangers.

6 *April*, 1674.—Which day the Deacon, Auditors, and whole Brethren of the Glover calling of Perth being met in their ordinary Meeting House anent the affairs of the Incorporation and taking to their serious consideration the abuse done by some of the freemen in colouring of Gloves to strangers, therefore they unanimously enact, statute, and ordain that none that whiten Gloves in the Calling shall upon any pretence whatever colour, whiten, or shew Gloves, garve skins, or make muffs, to any person whatever but to freemen of the Calling allenarly, as also that no freeman that sells Gloves shall take them again from those who bought them or from any others who gives them to whiten or Collour upon any accompt, the contraveener either Freemen or fials to be fined in Twenty pounds Scotts money *toties quoties* to be uplifted by the Deacon.

Leather to be washen before made into Gloves.

16 *July*, 1681.—Which day the Deacon, Auditors, and whole remanent brethren of the Glover Calling of Perth, being mett in a General Court within their ordinary Meeting House anent affairs belonging to the said Calling and taking to their serious consideration the several complaints given in by several Noblemen and Ladies anent our unwashen leather, how they have been abused thereby in time past Therefore the whole Court Do Enact, Statute, and Ordain for the future that no leather be wrought or almed [alumed] but that it shall be salted and dyed and then the salt washen out of it, and Ordains the Contraveener to be fined in Twenty pounds Scotts unforgiven *toties quoties*, to be uplifted by the Deacon and disposed of with advice of the Council of the Calling.

No Gloves to be made of tard forchets*, but with cliven forchets.

The Calling also enacts that no Gloves be made with tard forchets but allenerly with Cliven forchets under the Penalty of Fourty shillings Scotts, *toties quoties* unforgiven.

---

* The "forchets" in a glove are the inner sides of the fingers. If a glove is examined it will be seen to be cut out all in one piece with the exception of the thumb and these "forchets," for which the remaining fragments of the skin are used.

Gloves to be made wholly of one kind of Leather.

8 *October*, 1692.—Which day the Deacon and remanent Brethren of the Glover Incorporation of Perth convened within their ordinary Meeting House considering the great abuse done to the Calling by some of the Members making wild leather gloves with sheep leather forchets and thumbs, they therefore statute and ordain that in all time coming no Member of the Incorporation shall make any kind of Gloves but which he wholly finishes of one kind of leather, under pain of confiscation of such Gloves as are otherways made, and the transgressor to submit himself to the will of the Calling.

Freemen allowed to take an apprentice and employ him in the Science practised by himself.

10 *October*, 1724.—In a General Meeting of the Glover Incorporation of Perth convened within their Hall, they unanimously agree, statute, and ordain that any Freeman of the Incorporation may take an apprentice, and employ him in the particular Science known and practised by himself, notwithstanding of any Act made by the Calling to the contrary in the time past.

Act against selling Gloves in Dundee off Market days.

There was produced a Letter from the present Deacon in Dundee, Representing that some of our Incorporation had come to Dundee, and gone from Shop to Shop selling their Gloves, not upon public Market Days, to the prejudice of their trade and priviledges, which the Meeting taking into consideration, they unanimously Statute and Ordain that, if any Freeman shall be guilty of the like practice off Market Days they shall for each transgression pay one hundred pounds Scotts, for the use of the poor of the Calling, unforgiven.

Act against absents from Court.

5 *October*, 1754.—The Calling taking into consideration that there is Acts for imposing fines upon those who willfully absent themselves from Courts, they unanimously agreed to put the same in execution, and did further Enact, Statute, and Ordain that if any of the Bretheren of this Incorporation shall be guilty of absenting himself from Courts after being duly warned by Officer, without giving a reasonable excuse, and particularly if they are absent from the Court on the Saturday immediately after the election, when met to read over the Acts, then and in that case they shall be deprived of having a vote for a year to come, or pay Three pounds Scotts of Fine for the use of the Calling.

Act allowing to buy Leather or Gloves from Strangers.

The Calling considering that Act dated 9th November 1627, prohibiting Brethren from buying leather or Gloves from Strangers, and finding it detrimental, Did all agree to rescind and annull the same, and Enacts that any Member of the Calling may buy drest leather or Gloves from any Stranger, Skinner or Glover, providing always that they be well made, of good leather, and good and every way sufficient in their kind.

Apprentices to pay of Booking Money £3 Sterling. 3 *March*, 1763.—In a general Meeting of the Glover Incorporation of Perth convened within their Hall, The Calling agreed by plurality of votes to make an alteration in the booking money of Apprentices in future, and hereby Enacts, Statutes, and Ordains that from this time forth every apprentice bound to the Glover trade and Community thereof, in view of their having a title to enter freemen of the said Incorporation, shall pay to the Boxmaster for behoof of the Calling, three pounds sterling of booking money at their entry as an apprentice and also to be bound no less than four years as formerly, but allows any Member to teach whom they please to sew Gloves.

Deacon's Gloves. 3 *October*, 1810.—It was stated to the Meeting that by immemorial usuage every freeman at his entry had to deliver to the then Deacon a pair of Gloves, or in lieu thereof to pay a certain sum as the price or value of such Gloves, but as no sum was taxed as the value thereof nor any time fixed for paying the same it often proved difficult to recover the same or any adequate sum. Which being considered by the Meeting they Enact that in future all freemen shall at the time of their entry pay to the Deacon the sum of Five shillings Sterling in name of his gloves.

Gloves appear in the fourteenth century as exported from England, affording further proof that the home trade had become of some consequence. Anderson, in his *History of Commerce*, quotes from Sir Robert Cotton's *Records of the Tower*, an Act of 1378: "That all merchants, Gascoyne and English, may freely transport into Gascony and Brest, to the king's friends, all manner of corn and other victuals, and also leather gloves, purses, caps, and points (tagged laces), with other small merchandise;" and, in 1382, the Pope's collector of his dues in England obtained leave of Richard II. to export from Bristol, free of duty, a large quantity of apparel and furniture, among which appear "one capellum, and one pair of gloves lined with grey (marten skin); one pair of beaver gloves" (*Fœdera*).

In 1464 "arms" were granted to the Glovers.* This was part of a general movement by the trading companies for incorporation. The Drapers were granted letters patent in 1439, the Fishmongers in 1433, the Ironmongers and Haberdashers in 1407, the Taylors and Linnen Armourers (Merchant Taylors, as we call them) in 1466, the Clothworkers in 1482, with other less important bodies; while those even then well-established, as the Grocers, Goldsmiths, Skinners, and others, secured the confirmation of their charters. In 1483, we find an Act passed (1 *Rich. III.*) on the representations of glovers with other artificers of London, in which gloves, with a number of other small wares, were prohibited to be imported. The united petitioners declared that "in times past they were wont to be greatly employed in their said crafts for the sustenance of themselves and their families, and of many others of the king's subjects (depending on them), but of late by merchant strangers, denizens, and others, there are imported from beyond sea, and sold in this realm as much wares as may be wrought by the above-named artificers, now like to be undone for want of occupation." The crafts associated with the glovers in this movement were the girdlers, point-makers, pinners, pursers, joiners, painters, card-makers, wiremongers, weavers, horners, bottle-makers, and coppersmiths; and the enumeration of the wares manufactured by them not only makes a lengthy

---

* Technically described thus:—Per fesse sable and argent, a pale counter-charged, three rams salient of the second, two and one, armed and inguled or.—*Crest:* on a wreath, a ram's head argent, issuing from a basket of the last, between two wings, expanded gules.

catalogue of minor articles, but gives a more favourable idea of our commercial progress at this period than would be generally assumed. It should be pointed out that querulous preambles to ancient statutes are not always to be implicitly received. In this case, the desired end was gained, and, so far as gloves were concerned, a policy of protection began which lasted until the present century, always excepting the efforts of the " smuggler bold" to show the impolicy of prohibitions and the disadvantages of heavy duties. This Act of Richard III. only re-enforced a similar measure of 1463, which forbade "certain merchandise to be brought into this Realm ready wrought," and it was in turn re-enacted by a statute of the fifth year of Elizabeth, which prohibited imported gloves, with other articles, " to be sold, bartered, or exchanged in this Realm or Wales, in pain to forfeit the same or the value thereof, to be divided betwixt the Queen and the Prosecutor."

In 1479, we find mention of the glovers at Shrewsbury. The trade was followed also in the sister country, for in the red book of Kilkenny there is, in the sixth year of Henry VII. (1491), a warm dispute between the glovers and shoemakers as to the right of making girdles. In 1573, "the Aldermen, Comburgesses, and all the burgesses and coialtie" of Stamford, ordered that glovers were to pay for their absolute freedom—that is, at the expiry of their apprenticeship—a fine of 6s. 8d., or 10s. An Act of 1563, which provides that servants in certain employments shall be hired by the year, includes "gloovers ;" and Thomas Powell, in his *Art of Thriving* (a tract published in 1635, and reprinted in the fine Somers collection), in enumerating trades " that have in them some art or science, by which a man may live and be a welcome guest to all countries abroad, and have employment in the most stormy times at home, when merchants and shopkeepers are out of use," includes "imbroyderers," tailors, shoemakers, glovers, perfumers, and trimmers of gloves. Shakespeare, in the *Merry Wives of Windsor*, compares a great round beard to "a glover's paring knife," using what was, to him, a familiar illustration, since we know John Shakespeare, the father of our national poet, to have been a glover by trade, although, after his fortunes were mended by marriage, his original occupation is hidden under more extensive transactions in the brokerage of corn, and, more especially, wool. Still, Mr. Halliwell-Phillips inclines to the opinion that John Shakespeare continued to maintain his glove-making trade, only adding to it branches of business akin to it—buying sheep, stapling and selling the wool, and dressing the skins to be used up in making gloves. Instances have already been given of the purchase of raw sheep-skins by glovers, and it is not to be doubted that the first combination of the trades of fellmonger and tawyer with that of a glover long continued to be maintained. From a MS. dated 1595 it appears that some of John Shakespeare's contemporaries, who were glovers, also dealt in wool and yarn, besides generally following the trade of malting.

George Perrye, besides is glovers trade, usethe buyinge and sellinge of woll and yarne, and makinge of malte.

Roberte Butter, besides his glovers occupation, usethe makinge of malt.

Rychard Castell, Rother Market, usethe his glovers occupacion; his wieffe utterethe weekelye by bruynge ij strikes of malte.

And the Records of Rye, in 1604, contain an entry "To Townsen, the glover, for two sheepe skines, vj^s viij^d." In the churchyard of Stratford-on-Avon a tombstone, of the latter part of the seventeenth century, is erected to the memory of "Richard Smith, fellmonger and glover." January, 1688-89. (*Outlines of the Life of Shakespeare.*)

In 1638, the Glovers, under the title of The Master, Wardens, Assistants, and Fellowship of the Worshipful Company of Glovers of the City of London, obtained a Charter of Incorporation from Charles I. The original charter is not now in existence, and is supposed to have been destroyed in the Great Fire. A copy is preserved in the Guildhall Library. The preamble runs :—

Charles, by the grace of God, King of England, Scotland, France, and Ireland, Defender of the Faith.

To all to whom these presents shall come greeting.

Whereas, by an humble petition presented unto us by our loveing subjects living in and about our Cities of London and Westminster, useing the arte, trade, or mistery of Glovers,

Wee have been informed that their families are about four hundred in number, and upon them depending above three thousand of our subjects, who are much decayed and impoverished by reason of the great confluence of persons of the same arte, trade, or mistery into our said Cities of London and Westminster, from all partes of our kingdome of England and dominion of Wales, that, for the most parte, have scarcely served any time thereunto, working of gloves in chambers and corners, and taking apprentices under them, many in number, as well women as men, that become burdensome to the parishes wherein they inhabit, and are a disordered multitude, living without proper government, and making naughtie and deceitful gloves. And that our subjects aforesaid that lawfully and honestly use the said arte, trade, or mistery, are, by these means, not only prejudiced at home, but the reputation the English glover had in foreign partes, where they were a great commoditie, and held in good esteem, is much impaired. And, also, that by the engrossing of leather into few men's hands, our said subjects are forced to buye bad leather at excessive rates, to their further impoverishment.

Therefore the Company were, in addition to their prescriptive privileges, given the power to search for and

destroy all bad or defective skins, leather, or gloves. They
were empowered to have sixteen Assistants or more, four
Wardens, and a Master. The first Master was William
Smart, of St. Giles, Cripplegate—a parish which, according
to Hull, was then the centre of the London trade. The
fine upon admission was, at first, £5 13s. 4d. ; the fees at
present charged by the Company, which still maintains a
feeble existence, are, upon taking up the freedom by patri-
mony or servitude, £3 ; by purchase, £6 6s. 6d. ; upon
admission to the livery, £10 10s. ; upon election to the
Court, £3.

The " *City Press*" *Directory* states that by the bye-laws
of the Company (approved by the Court of Mayor and
Aldermen in 1680, and allowed by the Lord Chancellor
and Chief Justices in the following year), females are permitted to become members of the Company, and that six
were so admitted between 1780 and 1802. These were
early days for women's rights and wrongs to be recognized.

There is now no Court of Assistants in the Company,
and apparently need of none. The liverymen appearing
as voters for the City, in 1881, only number fourteen in all,
including the Master. The Company has, what is more to
the purpose, no estates, but exists on an invested capital of
£1,800 and £1,250 Three per Cents. The dividends
upon the smaller sum are distributed in pensions to liverymen and freemen who can plead poverty, or to their
widows.

The Perth Incorporation, although flourishing enough,
also represents an extinct trade. At one time, Perth gloves
had a wide reputation, and were extensively exported, to
say nothing of between 2,000 and 3,000 pairs annually

produced for home use. In 1795, the lamb and sheep skins used in the manufacture amounted to over 30,000. The Skinner gate was almost wholly taken up by the shops of thriving glovers; and one, Bailie Gray, alone employed seventeen men cutting gloves to keep his sewers in work. The glovers made buckskin breeches also, and their usual sign was a pair of breeches with a buck between the legs, to show the combined manufacture. Over English glove shops, it was customary to hang out a huge hand. By 1836, the last operative glover in Perth had been gathered to his fathers, and what was left of the once flourishing trade had been taken to Dundee by the son of an old Perth glover. The calling at that time numbered 64 persons, and had an income, in 1832, of £1,094 4s. 8½d. and an expenditure of £960, of which the eleemosynary roll took up £498 10s. 9d. On its "poor's roll" were 21 persons, and it incurred charges for gifts of coals, shoes, and clothing, as well as grants for education, besides having then recently adopted a scheme for the payment of annuities to its superannuated members and widows. The fees on admission were—

	£
Sons of freemen	1
Apprentices, under 30 years	20
„ „ 40 and under 50 years	27
„ „ 50 and upwards ...	50
Strangers, who must be operators	100

The Incorporation is yet represented in the City Council and has an annual dinner at Michaelmas.

M

CHAPTER XII.

The Glove Trade.

PROHIBITION of the import of foreign-made gloves into this country, as enacted by statutes of Edward IV., Richard I., and Elizabeth, continued in force until the present century. These measures were by no means exceptional, but only represented the prevailing spirit of legislation. To protect yet more stringently the home trade, another Act was passed in the sixth year of George III. " for the encouragement of the importation of foreign kid and lamb skins unmanufactured," which provided that any foreign manufactured leather gloves or mitts might be searched for by any customs or excise officer, and when found seized. Any principal or accessory in the act of importation, any vendor or person having possession of such gloves, as well as any who might endeavour to conceal them, should forfeit the sum of £200, and pay double costs. The seized goods were to be sold " by the candle," a mode of auction by which people were allowed to bid during the burning of an inch of candle, the goods being knocked down after the light had gone out to the last bidder. The proceeds of the sale were to be divided in equal portions between the officer who made the seizure and the crown.

The times called for this renewed imposition of old bonds. The former laws had become as so many whips locked away in a cupboard, to be brought out and flourished in the face of the people only as occasion required. Time dulled vigilance, customs officers became less curious and more complaisant, and the door of trade was pushed further and yet a little further open until the abominable thing was freely admitted, and people could at will buy and wear gloves which had, in an English glover's eyes, "the mark of the beast" upon them. If the unlawful trade became too brisk, the old laws were remembered and revived, and for a time all was well again, until quiet and stealthy encroachments again began. But when this Act of George III. was passed, another danger to the tender English trade, coddled and nursed under prohibition, had, in the spread of smuggling, assumed large proportions, although it was not by any means a novelty or new evil. In a tabular statement, drawn up in obedience to the commands of "the Right Honourable the Lords Commissioners for the Treaty of Commerce with France," and "humbly tendered to their Lordships," in 1674, setting forth the particulars of the trade between the two countries, " Jessamin Gloves " appear among the " Toys for Women and Children, Fans, Laces, Point Laces, rich embroidered garments, beds, and other vestments, which are of an incredible value," imported from France. This scheme, as it is called, published in Lord Somers' *Collection of Tracts*, more than hints that many of these toys are "conveyed away privately." From this point, contraband trade had increased in exact ratio to taxation, until it had become, early in the eighteenth century, a danger to the revenue, demanding constant attention from the legislature, and a demoralising influence of undefinable extent. Laughing at laws, and particularly at those which demanded duties, the smuggler had taken trade into his own hands, caring nothing about home manufactures or national revenue so long as he could evade or defy the " Revenue Sharks," and carry on his own business. Smuggling had become a recognized and lucrative calling, and was gradually passing from the small skipper of a swift lugger or schooner, into the hands of capitalists, who owned sometimes several vessels built expressly for "the Company," well armed, and worked by a regular staff of trusty, resolute men, ready to take any risk and having a particular and personal hatred of gaugers and revenue men. Was a cargo of brandy, or silks, or tea, required to be landed, or a small parcel of laces and gloves, wanted by a customer?—the payment of an insurance premium at a well-known office abroad would ensure safe delivery. " The smugglers," it was said in the House, in 1831, " could insure the delivery of French goods into any warehouse in London, at a charge of nine per cent. upon their value." A witness named M'Gregor, examined before a Parliamentary Committee appointed to consider the import duties generally, said that the smuggling of gloves could be carried on for about nine per cent. upon their cost. At this time the official duty ranged from 20 to 40 per cent upon the value of the gloves brought in. These high duties practically protected the smuggler, besides remunerating the risk he ran. They ensured popular sympathy with " the Master in the Private Line." One of Sydney Smith's essays in the *Edinburgh Review* gave the humorous side of the grievous burden of taxation under which the English citizen strove to make headway. Taxes dogged all his steps through life. In childhood he played with a taxed top; in later years he drove a taxed horse by a taxed bridle along a taxed road; in his last illness he took medicine, which had paid 7 per cent. duty, from a spoon which had paid 15 per cent., under the direction of an apothecary who had paid a license of a hundred pounds for the privilege of putting him to death; his property was then charged with probate, fees were levied upon his burial, and he was laid to rest, if he could, under taxed marble. This only made light of a sore infliction from which possibly none but the reverend humorist could have extracted fun. It was not to be wondered at that smugglers throve.

People are, in the main, often averse to putting the proper construction, from the official point of view, upon a fiscal duty. Their duty, they consider, is to pay as little as possible; and at this time there was a very general leaning towards cheap brandy or tobacco; "old gown," as smuggled tea was called, was in general request, and there was thought to be no wrong in buying, if possible, contraband silks, laces, and gloves, which were excellent in quality and comparatively cheap. Thus smuggling appeared as a romantic struggle with constituted (and exacting) authority, and there was roused a keen and not inactive interest in the "owler," as these night birds were known. As far as possible the people protected the smuggler by giving him timely warning of an official ambuscade, or by offering the officials confidential—and misleading—information. Often the populace collected about a landed cargo in such numbers, and were so threatening in their aspect, plainly lacking neither the ability nor the will to take active part in the proceedings, that the revenue officials, and the military called in to their aid, were reduced to quiet spectators of the affair. This is affirmed by the Report of a Parliamentary Committee, which sat in 1783, to enquire into the evil. Some fifty years before then, Sir Robert Walpole had declared his belief that in most of the maritime counties the military would, of the two, protect the smuggler rather than the exciseman; and in the House of Lords, about the same time, it was stated that, on the coasts, the whole people were so engaged in the traffic that it was impossible to find a jury that would, upon trial, do justice to an officer of the revenue. Even if precautions and signals could not prevent a collision, and some of the poor culprits happened to be captured, escapes were not infrequent; nay, even if they were brought before his worship, the neighbouring justice of the peace, it often happened that some legal loop-hole was discovered, or an arbitrary exercise of judicial stupidity made a way to set the smuggler free. It was too often even to his worship's interest to be either captious or dense, unless his worship was sadly belied, for he too loved, not wisely, but too well, some good claret or brandy, and Mistress Shallow had gowns and gloves by no means above suspicion.

Until 1766, the penalty attending detection in the offence of bringing in foreign gloves, beyond the personal necessities of the offender, had been £20; and that, on recovery, was equally divided between the person at whose instance the law had been put into force, and the poor of the parish. Holt's *Dictionary* of 1756, taking the very charitable view that many persons had innocently and inadvertently engaged in encouragement of illicit trade, by trafficking in contraband gloves, " by which great quantities of French and other foreign-made gloves are brought into the kingdom, and sold, contrary to law, to the prejudice of His Majesty's revenue, and the detriment of the glovers and all others concerned in the manufacture of gloves," ventures to specify the manner in which an offence could be committed.

Every person offering such gloves to sale forfeits the same, with treble the value, and the gloves may be seized by the person to whom they are offered. Every person buying such gloves forfeits the same, with treble the value, and the gloves may be seized by the seller ; and either buyer or seller may prosecute the other, and the first prosecutor be acquitted. Every person harbouring or concealing such gloves forfeits the same, with treble the value. Every person receiving or buying such gloves, upon conviction before any Justice of peace, on the oath of one witness, to a fine of Twenty Pounds.

It would appear, too, that in the time at which this work was published, a trade in fictitious smuggled goods was carried on, even as in our own day one may be accosted at a seaport by an elaborate sailor, with a request to buy "a bit of good tobacco," which when consumed does not add to the reputation of foreign countries,, or conduce to the consumer's comfort. "Many gloves," says this writer, "are sold under a pretence that they have been seized and condemned at the Court of Exchequer, and then legally sold. It is necessary to inform the public, that all such gloves as are really condemned and sold in the Court of Exchequer, are stamped at the custom-house before they are delivered, with an oval stamp between the letters G. R., in which is a crown ; above it, these words, in capital letters—' Custom House;' and below it, ' Seized,' as a mark to distinguish them as having been condemned and sold in the said court; and consequently such gloves, which are not so stamped, cannot be legally bought or sold."

The revenue on gloves was often defrauded, without

running any danger, by a device which had all the merit of genius, and yet was simple to a degree. The gloves were packed up in two loads of separate consignments ; the one batch consisting all of gloves for the right hand, the other all for the left. If both lots were successfully run, well and good—a satisfactory re-union was effected between the separated pairs ; but if one lot fell into the hands of the Custom House officers, and were in due time put up for sale, the possessors of the left-hand gloves would be able to buy up the right-hand ones that had been captured, and *vice versa*, at the merest trifle—care having been taken to let it be widely known beforehand that the gloves offered in the packages were all for one hand only. (*Reminiscences of an old Draper*). If the articles offered did not realize the amount of the customs' duty, they were burnt in a huge kiln, still known as "the Queen's tobacco pipe." On one occasion 45,000 pairs of gloves were consumed. The indignation properly felt at so wanton destruction may be tempered by the reflection that the consignee had probably taken out previously some small parcel of the gloves, and, finding them to be damaged in transit, had refused to pay heavy duties upon worthless articles, and so increase his loss.

The duties once in force are shown in "A New View of the British Customs," included in Malachi Postlethwayt's *Universal Dictionary of Trade and Commerce*, 1755. Although tariff tables are perhaps the least interesting of anything concerned in "the dismal science," as political economy has been styled, yet that which follows will well repay study, not only as showing the descriptions of gloves formerly imported, but because it shows the incidence of taxation.

GLOVES.	Rates of Car. II. and Geo. I.			Paid on Importation								Repaid or drawn back on exportation in time.			
				By British.				By Strangers.							
	£	s.	d.	£	s.	d.	1/100	£	s.	d.	1/100	£	s.	d.	1/100
Of Bruges making, the groce containing 12 dozen	2	10	0	0	11	11	62½	0	12	7	12½	0	10	9	37½
Of Canary, Milan, or Venice unwrought, the doz. pair	1	0	0	0	4	9	45	0	5	0	45	0	4	3	75
Wrought with Gold or Silver, the doz. pair	4	0	0	0	19	1	88	1	0	1	80	0	17	3	…
French making, the groce containing 12 doz.	2	10	0	1	13	10	12½	1	14	5	62½	1	0	1	87½
French, wrought with Gold or Silver the doz. pair	4	0	0	2	14	1	80	2	15	1	80	1	12	3	…
Spanish, plain the doz. pair	0	15	0	0	3	7	8¾	0	3	9	35¾	0	3	2	81¼
Of Vandon	0	10	0	0	2	4	72½	0	2	6	22½	0	2	1	87½
And besides, if the leather of the aforesaid Gloves be the most valuable part, for every 20/- of their real value upon oath	…	…	…	0	6	0	…	0	6	0	…	…	…	…	…
Of Silk, knit, the doz. pair	2	0	0	0	11	5	40	0	11	11	40	0	10	6	…
And besides, for every pound weight	…	.	…	0	1	10	50	0	1	10	50	0	1	10	50
But Gloves may not be imported by any person to be sold															
Glovers' clippings, the 112 lb.	0	5	0	0	1	2	36¼	0	1	3	111¼	0	1	0	93¾

These rates set forth the presumed value of the gloves brought in. Those sent out of the country were of "sheep, kid, or lamb's leather," "fringed and stitched with silk," "furr'd with Coney (rabbit) wooll," and those "of Bucks' leather," which range in value from 4s. to 20s. per dozen pairs. A Charter purchased by the City from the King in the sixteenth year of Charles I., gives tables of the Rates of Scavage and Paccage to be charged by the Corporation. The Scavage rate was imposed on imported goods for the privilege of exposing them for sale, and here we find 2d. per dozen charged upon "Gloves of Spanish leather." The Paccage rates charged on exported goods were :—

	s.	d.
Gloves, Bucks' leather, the Dozen Pair	1	0
Gloves, with Silk Fringe and Faced with Taffety, the Doz. Pair ...	1	0
Gloves, lined with Coney or Lamb skins, or Plain, the Dozen Pair	0	2

The *Book of Rates* of Charles II., on which Postlethwayt founded his information, was the first collection of the Customs duties into a convenient form, and this continued in force until amended in the reign of George I. The rates set forth in these tables on which percentages were charged, constituted the "poundage" on merchandise, which is so often coupled with the "tunnage" on liquors in our records, and made up the Customs revenue. But beyond these, which were necessarily untrustworthy as values fluctuated, there were so many variations of the tariff, surplus charges of so much per cent. on specified articles, so many new methods of arriving at charges, and exceptions in favour of certain countries or their products, that it is a matter of no small difficulty to trace the impost on any article of commerce through successive years. At one time gloves, after they were, in 1825, permitted to be freely imported, were charged with an *ad valorem* duty, ranging from 20 to 40 per cent. In 1844, "habit gloves and mitts" were charged with a specific rate of 2s. 4d. per dozen pairs ; "habit gloves and men's gloves," 3s. 6d. ; and "women's gloves or mitts," 4s. 6d. per dozen, as reduced from 4s., 5s., and 7s. per dozen respectively for the same varieties in 1835. The duty on gloves was wholly repealed in 1860, when more than six millions of pairs were imported, producing a revenue of over £65,000. We bring in now over thirteen millions of pairs, valued officially at above £1,300,000.

In 1785 a tax was imposed upon gloves. New taxes were not novelties about that time. Every Budget brought a fresh relay of experimental imports, or of additions to those already established, but not yet brought up to exhaustion point. In the year just mentioned, for instance, in addition to the tax on gloves, others were put upon servants, both male and female, upon shops according to their annual value, upon attornies, regardless of their value, upon post-horses, and pawnbrokers, and, by a new regulation, upon salt. In one respect or another the preceding financial year placed under contribution candles, bricks, hats, pleasure horses, British linens and cottons, ribbons, liquor-retailers, paper, hackney-coaches, silver plate, lead exported, silk brought in, and postage, with the right of shooting game ; and, again, in 1783, besides stamp duties in considerable variety, including the familiar stamp on receipts, there were taxes laid upon turnpike road and enclosure bills, upon entries of births, christenings, marriages, and deaths, upon medicines and vendors thereof, upon stage coaches, diligences, and waggons. An average of ten new taxes per annum would severely try the patience of the people in these tariff-reducing times, but a long course of political phlebotomy, made necessary by the extensive and exhausting wars with half Europe, and with the revolted Colonies to boot, had made the British taxpayer in those days very tractable. In introducing the Budget in 1785 " the heaven-born Chancellor," Pitt, said that a tax upon gloves was one that had attracted the attention of many previous Chancellors of the Exchequer, but a difficulty had always presented itself in finding a mode of collection which would allow of the tax being made a profitable one. The experiment tried the year before with hats, of affixing a stamp to each one sold, had suggested that the same method might be employed with gloves, the stamp to vary according to the value of the gloves. " By that means the tax was collected from the retail trader, and he had no occasion to lie out of his money before the sale of his goods, which was a reason why he need not make an additional advance on his customer, as was generally the case where the tradesman had a large part of his capital engaged in paying duties that were not repaid him for a considerable time." In arriving at the probable yield of the tax, it was necessary to estimate the number of persons who might be supposed to wear gloves. These were placed at three millions. Each of them, it was presumed, would wear at least one pair of gloves per year, although some, it was thought, wore indeed twenty, thirty, or even forty pairs in that time, but, not to be too sanguine, the average wear of the whole was to be put at three pairs per annum for each person. On these, where the value of gloves was between 4d. and 10d. a pair, a tax of 1d. would be put ; from 10d. to 14d. a pair, 2d. ; and on all gloves above 1s. 4d. in value, a tax of 3d. a pair ; and this, Pitt estimated, would bring in a revenue of £50,000 per annum, a calculation, both in detail and result, which Fox thought greatly over-estimated. The Bill also imposed a license upon the retailer of gloves. The proceedings of the House, when the Bill was considered in Committee, are reported in the *Gentleman's Magazine* (*vol.* lv.) in a manner suggestive of a well-known nursery legend, even though parliamentary reporting, practically begun by Cave in this fine old publication, was then nearly fifty years established.

The House resolved itself into a committee on the glove tax.

Mr. Coke wished to know if it was intended to extend to silk mittens and gloves ; in which case, added to the heavy duties on the raw materials, the manufacturers would be most grievously loaded.

Mr. Rose replied that it was meant to extend to all sorts of gloves ; nor could it possibly be a grievance to any in particular, as the tax would be general.

Mr. Smith (for Worcester) proposed a clause to guard the English manufacturer from injury, by obliging dealers, on renewing their annual license, to swear that they had not disposed, during the preceding year, of gloves of foreign manufacture, or gloves that had not paid duty.

Mr. Eden opposed this idea, unless made general, and that dealers of every kind should swear the same.

Sir J. Johnstone said, the clause would ruin one half of the dealers, and damn the other half.

Mr. Pitt and Mr. Rose thought the clause a good one.

Mr. Attorney-General spoke forcibly against it.

The clause was withdrawn, and the bill agreed to.

Nine years later (36 *Geo. III. c.* 10), the duty, excepting the license on the retailers, was repealed, because, as Pitt admitted, it produced so little and was found so vexatious. Again, two years after (38 *Geo. III. c.* 80), the licenses also ceased and determined, and the home glove-trade was left free and unshackled; always excepting the competition with foreign manufacturers, brought about by the smugglers—a competition constantly being checked, as far as preventive measures and preventive men could compass that end.

Freedom was what the glovers least desired. Prohibition gave them a practical monopoly of the home market, and there was a progressive prosperity in our commerce which even wholesale wars could not wholly check. In this sleepy sunshine the glovers basked content, and it was a very rough rousing that they experienced when their trade was brought under the Free Trade legislation, inaugurated in 1823 by Huskisson. In 1826 all restrictions upon the importation of foreign-manufactured gloves were abolished, and the English manufacturer was placed in direct competition with his foreign rival, who was, however, made subject to specific duties on bringing over his goods. The result was disastrous. The glovers saw that they must bestir themselves, and the new tariff having considerably reduced the import duty on skins used in the glove manufacture, they brought in the raw material in largely increased quantities, and set vigorously to work to make it up. For a time, there was a wonderful increase of business, production was stimulated to the utmost, and masters and men must have congratulated themselves on a great good having come out of an apparent evil of undefinable magnitude. People flocked into the glove-making centres to share in the plethora of work and wages. But the work was carried on in the same manner and with the same output as aforetime; no new market or increased demand was found to carry off the flood of production, stocks accumulated on the hands of the masters, and dire distress set in. The output was first limited by partial employment of the workpeople, then by their discharge on a large scale. Deprived of their livelihood these "went on the rates" or entered the poor's-house; the burthen of local rates increased enormously, and, by reason of the new population which had come in, to an inordinate degree—creating a fresh evil where surely none was needed. Many of the distressed operatives, who had been earning from 20s. to 30s. a week, were reduced to stone-breaking and road-mending at "8d. a day; women, 4d." It was resolved to ask Parliament for a reversal of the new policy and for a return to the prohibitory system. Petitions were drawn up and signed, and presented by local members, and in both Houses no opportunity was lost of raising a debate on the general question. Failing a direct measure of intervention, of which no hope was held out by the ministry, motions were made in the lower House by Colonel Davies, and in the House of Lords, by Viscount Strangford, praying that a committee might be appointed to inquire into the state of the glove trade. It was pointed out, and with truth, that the glovers had always been industrious and independent, and that while many other classes of artisans were leaning towards sedition, and coquetting with agitators of the worst type, these had ever remained peaceable and orderly. The manufacture itself was pictured as being almost idyllic; carried on by the operatives, men as well as women, in the quiet and retirement of their own homes, away from all the over-crowding and, as then held, immoral and unwholesome influences of the newly introduced factory system, masters and workpeople were shown as working in undisturbed harmony and goodwill for their common benefit. That poverty had intruded upon these peaceful scenes and happy homes, that the masters had been brought to the verge of ruin, was attributed wholly to the new Free Trade system, to the increased import of foreign gloves, and the still vigorous smuggling trade.

These statements were all directly controverted by the other side, or directly contrary constructions put upon them. The glovers, it was said, had only themselves to thank for the misery they were experiencing; as soon as trade recovered from the depression, caused by reckless over-production, all would be well again with them. There was no increase in the importation of foreign gloves, but the newly-introduced cotton " Berlin " gloves—such as coachmen affect nowadays—had been so generally adopted as to materially affect the demand for skin gloves. One member improved this argument by holding up in the House a pair of " Berlin " covered hands. To this it was replied that many of these self-same " Berlin " gloves were actually made in Yeovil and other glove-making centres, so that distress could not have come through them.

Still it was urged that no relief could be given to the glove trade at the expense of all other industries; depression was general, and one and all must wait until the tide of trade turned. Was this general depression due to Free Trade ? Certainly not, said the ministry. And as for smuggling, there was no smuggling to speak of. The Chancellor of the Exchequer declared that only one hundred dozens of contraband gloves had been brought in during the whole of the previous year. Lord Auckland, the President of the Board of Trade, on the joint testimony of smugglers and custom house officers, was prepared to say that no illicit trade in gloves was carried on at all. This statement led to one of the most animated passages in a debate not more tedious than ordinary. Viscount Strangford, who led the question in the House of Lords with much spirit, showed that such a quantity of gloves, one hundred dozen, could be packed and compressed in a case eighteen inches square—not small women's gloves, but gloves that would fit " the fullest-sized Peer· or personage in that House." Was it to be believed that only one such very portable, and, to the smuggler, very convenient, case was smuggled into the United Kingdom in the course of a year? And, said Viscount Strangford, as to the conjoint testimony of custom house officers and smugglers, adduced by Lord Auckland, the one would say that there was no contraband trade to prove his vigilance, and the other to declare his innocence, much after the manner of the old Spanish fable, where the cat bore witness to the honesty of the rat, and the rat testified to the vigilance of the cat. The noble Viscount went on to declare that—

At the very moment of the noble lord's speech, he made a vow that he would deprive him of the triumph which his confidential intercourse with smugglers had given him. So he, too, proceeded to place himself in communication with smugglers—

" Et nos quoque in Arcadia viximus;"

and he was proud to say that he succeeded in finding some as daring, and as desperate, and as reckless as any, or almost any, set of men with whom the noble lord had ever associated—and he thought he had associated with men of that description, daring, and desperate, and very reckless indeed ! Now, his smugglers said, in flat contradiction to the noble lord's smugglers, that smuggling had increased, was increasing, and ought to be—no, that they did not say—ought not to be diminished. Let the noble lord but grant him the Committee, and full protection for his contraband friends and for himself, and he would produce, and the noble lord should himself examine, them.

These debates were raised on several occasions, but invariably without effect. Ministers were not to be moved, and divisions always proved adverse. The glovers began to look for help elsewhere. Depression still continued,

N

and trade did not mend. Of the hard struggle for existence which was then experienced, many old operative glovers still working at Worcester can speak feelingly. The trade grew worse rather than better ; and about 1840 the distress had become very serious, so that a deputation waited upon the then Bishop of Worcester (Bishop Carr), at Hartlebury Castle, to ask him to present some gloves of Worcester manufacture to Queen Adelaide, and to raise a fund to relieve the most pressing need. Royal influence in fashion is, however, very transitory, and public charity is not perennial, however nobly appeals are at first responded to. What was far more to the purpose, the glovers at this time started strenuously to help themselves, improving their manufactures and pushing their trade. In the thin kid, France then had, and still maintains, unapproachable excellence ; but in the stouter skin gloves England stood, and still stands, pre-eminent. Forsaking all thoughts of competition, save in quality and excellence of workmanship, the glovers began to forge ahead, and from that time until now, the trade has enjoyed a happy monotony of prosperity. At the present time English gloves, of which the manufacture centres at Worcester, are without rivals, are exported to all parts of the world, and command higher prices than any other.

In some respects, indeed, the glove trade was peculiarly favoured. Towards the close of the last century began a forward mechanic movement which revolutionized production. All the old boundary lines of trade were broken up and blotted out by the successive inventions of men whose magnificent energy and fertile genius nothing could withstand. Crompton, Hargreaves, Paul, Kay, Highs, and Arkwright attacked cotton and subdued it, giving us new political power and the foremost place in the mercantile world. Cartwright gave to the loom a multifold value; Jacquard, to all seeming, endowed mechanism with mind; while Heathcote, rivalling the spider, produced by machinery a filmy web of bobbin net. The application of steam by Watt, Boulton, and Stephenson gave to all of these a willing and untiring servant, with a capacity for work which could not be exhausted. Later on, wool and silk were overcome, and, last of all, linen. But leather glove-making, a delicate and difficult operation, apparently defied the irruption of invention. Circular stockings could be knitted, as well as gloves of various threads, on looms which became more and more complex and adapted to meet successive difficulties. Several machines for sewing and cutting gloves were introduced, but for a time with very small encouragement of success ; and even when a French " punch," in the shape of a hand, cut out gloves by wholesale, supplementing an older "punch" of English invention, used for what are known as " fabric " gloves, and when the sewing machine was adapted to the ornamental stitching of gloves, the sewing of the seams, the operation which affords most general employment, still remained undisturbed. " It is," wrote William Hull, "a happy circumstance for the operative glovers that machinery cannot be brought into operation against them," and it seemed probable that glove-sewing, with lace-making and straw-plaiting, would pass into history, as the last relics of our once extensive domestic manufactures. Now, however, a machine of German construction produces gloves entirely by automatic power, saving only one minor and unimportant process, known as " felling the slit-welt"—that is, the turning over and hemming of the welt on the edge of the opening of the gloves. Thus the entire subjugation of gloves to mechanic power has been effected. These machines have, although not absolutely of recent introduction, now begun to gain ground, and will probably continue to do so, displacing hand-labour in proportion. At present, however, there is no fear of their entire supremacy, and the glove manufacture, although in presence of an enemy that may prove more and more aggressive, still retains its most pleasing features. The factory system, which so excited the ire of Viscount Strangford and his friends, obtains only partial hold upon it, and most of the work is done by the operatives at their own homes. All about Worcester, whole villages depend on the labour of deft-fingered women, to whom, periodically, parcels of unmade gloves are brought to be stitched by the agents of glove firms, and, in due time, again fetched away. The collecting vans of our premier glove-making firm—Messrs. Dent, Allcroft & Co., who employ some thousands of glove-sewers —cover a radius of some fifty miles around Worcester, but the industry extends to villages in Shropshire, Warwickshire, Oxfordshire, Gloucestershire, Dorsetshire, Somersetshire, Devonshire and Cornwall. The different stages through which the gloves pass before being ready for the wearer, as well as a knowledge of the incidence of this home manufacture, can be found in the pages of Mrs. Henry Wood's novel, entitled *Mrs. Haliburton's Troubles*, by those who desire to gain, without much trouble, an insight into the technicalites of glove production.

Of the materials employed in glove-making there is a very decided limit. For centuries after the introduction of gloves they were only made of skins, and principally from those of the lamb and deer, often, like the "dannocks," or hedging gloves still used by labourers, made up with the fur inwards for the sake of warmth. Gloves of cloth and knitted from thread and cotton of Nottingham manufacture, and silk and worsted from Derby and Leicester respectively, have at times interfered considerably with the skin trade, but without affecting the ultimate demand. The materials employed have been considerably restricted, as well as the ornaments and accessories supplementing them. Certainly, we hear at intervals of novelties in gloves, such as the addition of fringes or tassels ; of the insertion of gussets of lace or embroidery at the back or on the wrists ; of the wrists being edged with puffings or pinkings, or looped with ribbons ; and even sometimes of flowers, real or artificial, being used to set off the native grace of a well-gloved hand, but it cannot be said of any of these frivolities that they interfere with the demand for a good plain pliable glove, well cut and well made. Kid gloves are the best and most expensive, but are far more rarely met with than most people would credit. The majority of the gloves sold as kid are made from lambskin, those known as doe, buck, or dog-skin, from the skins of sheep or calves. Kid-skins are largely used in glove-making, the kids being reared in several European countries, to the great benefit of the peasantry, but nowhere so well as in France, where kid-culture is carried to perfection, and so sedulously practised as to be something more than a mere addition to ordinary duties. To this is due the value of the French skins, which command higher prices than any in the market. The kids are not allowed to roam at will, when they might injure their precious skins by pushing through prickly hedges, or rubbing against rocks and doorways, but are kept carefully confined under a coop. Here they are fed with milk only, again, for fear lest coarser food should give a corresponding quality to the skin, to its great detriment. This may seem coddling rather than care, but the result is that larger and more delicate, and consequently more valuable, skins are obtained when the kids reach the end of their short existence, than when they are allowed free pasturage and the indulgence of their playful or pugnacious proclivities. Still, the value for glove-making of kid-skins that are not reared with anything but ordinary care has led to the suggestion that the English labourer might supplement his scanty earnings by keeping a couple of goats, or more. John Ploughman will have nothing to do with bees, nor will he take up the rearing of poultry for their produce. He does not mind a pig, if he can afford to buy one, and will lean over the wall of the stye for hours in contemplation of his unwieldy favourite; but "them other things" have no attraction for him, nor can he be persuaded that there is much to be gained from them. Goats, however, requiring more control, and appealing more forcibly by their size and their fighting capacity to John's sympathies, might become a favourite with him, and a source of profit, too, for it would not be easy to exhaust the kid-skin market by anything short of wholesale production. A sense of the value of kid-skins in ministering to the needs of fashion led some enthusiastic and far-sighted individual, some time ago, to write to Senator Mackay, the well-known Nevada millionnaire, with a proposal that they should unite in buying up all the goats in the world, and so establish a monopoly of kid leather—a project worthy of the South Sea bubble period. Another proposal, more feasible, but not less a failure, was that of some French speculators, who thought to established a rat preserve in Chicago, the most rat-ridden city in the world, and so to supply Paris with rat-skins for conversion into "kid" gloves.

The mention of dog-skin gloves reminds us of an occasion on which they served to show most elaborate courtesy and declare supreme devotion. Antonio Perez, one time Spanish ambassador, sent to Lady Knolles a pair of gloves with a letter, saying, "These gloves, madam, are made of the skin of a dog, the animal most praised for its fidelity. Deign to allow me this praise, with a place in your good graces. And if I can be of no other use, my skin at least might serve to make gloves." This conceit so pleased its author, that, in a letter to Lady Rich, he again employs it, and with greater elaboration. "I have endured," he writes, "such affliction at not having ready at hand the dog-skin gloves desired by your ladyship, that I have resolved to sacrifice myself for your service, and to strip off a little skin from the most delicate part of myself, if, indeed, any delicate skin can be found on a thing so rustic as my person. . . . . The gloves are of dog-skin, madame; and yet they are of mine, for I hold myself a dog, and intreat your ladyship to hold me for such, as well on account of my faith as my passion.—The skinned dog (*perro decollado*) of your ladyship, Anton. Perez." Lady Rich might have easily tested the protestations of her effusive admirer, if there be anything reliable in the theory of instinctive antipathies and affections to which our forefathers pinned a part of their very comprehensive faith. We find it solemnly laid down, as a fact worthy of full credence, that a drum made of wolf-skin would dominate and break another made of sheep-skin, if they were placed near each other, and that hens would fly at the sound of a harp strung with catgut. So, Sir Kenelm Digby, in his treatise *Of*

Bodies, says, "We daily see that dogs will have an aversion to glovers that make their ware of dogs' skins; they will bark at and be churlish to them, and not endure to come near them, though they never saw them before."

Of the manner of gloves that our forefathers wore, we get a complete enumeration in the *Cyclopædia* of E. Chambers, published in 1741. Here, besides the ordinary wear of leathern, silk, thread, cotton, and worsted, of which the leather are divided into shammy (chamois), kid, lamb, doe, elk, and buff, there are also, "single" gloves and gloves lined, tip'd, laced, fringed with gold, silver, silk, and fur; gloves of velvet, satin, and taffety; and gloves perfumed, washed, glazed, and waxed. These glazed gloves are often met with in old bills, and had then been long in use. The Colonial series of the *Calendar of State Papers* show a report to the East India Company from Surat, of 12th March, 1619, stating that, "The sale of sword-blades, knives, glass-ware, ... and the like, yields little profit, but are fit for presents. Supplies of morse teeth, cochineal, pearls, enamel-gloves and bone-lace, may be refrained from altogether. Camlets, mohairs, or the like, from Turkey, not profitable, these people bringing them from Mocha much cheaper."

Gloves of jean, satteen, and other coarse fabrics, although once generally sold, have been altogether superseded by the cheap and warm gloves now at the service of the poor. Gloves have been made from horse-hides, but without any idea, as may be supposed, of their becoming popular. The Eskimos make rude gloves from walrus-hide, sewn with threads of sinews, and "there is a passage in Du Cange from which it would seem that whale-skin was sometimes employed." Gloves have been made from the fibres of nettles, plants which have been used for centuries, and in many countries, for making cloth, and which it is thought may yet afford the textile of the future. Antiseptic gloves have been made for medical men to provide a doubtful remedy against infection, and among the many portable articles that, left about on the deck of a ship, might save lives in case of foundering—hats, cloaks, cushions, belts, mattresses, umbrellas, and what not—gloves have been included. Some of these were on view in the World's Fair of 1851. Gloves have been made of asbestos fibre, so that, like the napkins mentioned by Pliny, they might be cleansed by merely casting them into the fire. Cottrell mentions them as made at Ekaterinburg (*Recollections of Siberia*). Gloves have been made, too, from the fibres of the byssus, the beard of the molluscous *pinna*, "the silk-worm of the sea," so sedulously sought by the Sicilian fishermen, and made up by them into various articles of great fineness and delicacy, and of price in proportion. A pair of gloves made from the byssus were presented to Pope Benedict XIV. And, as a last curiosity in glove commerce, they have been produced from spider-silk by M. Bon, a French naturalist, who, in the beginning of the eighteenth century, created a commotion in the scientific world by claiming the homely and only too easily propagated spider as a certainly successful rival to the delicate silk-worm. If spiders had not an unpleasant habit of consuming each other, something might have come of the researches of this enthusiastic philosopher, who would have found Laputa more friendly to his labours than France, for it was practically proved that silk could be produced from spiders, and in sufficient quantity to make small articles from. Some spider gloves were presented by M. Bon to the Royal Academy of Paris, and others to the Royal Society of London. Since, however, it took nearly seven hundred thousand spiders to produce a pound of silk, and every spider of them would eat as many other spiders as could be overcome, there have been no further manufactures of this nature, and there is very little hope of such fabrics coming into general use.

SYMBOLICAL.

———◆———

"CLOTHING THE PALPABLE AND FAMILIAR
WITH GOLDEN EXHALATIONS OF THE DAWN."

—Coleridge.

GLOVES.

CHAPTER I.

Gloves as Pledges.

TO read *The Fair Maid of Perth* is to have a liberal education in the art of glove-making as practised by honest Simon Glover and his fellow-labourers in the days of old. Scott could have written a History of Gloves with ease; he had plainly given much study to their antiquity, and availed himself freely of his knowledge in this stirring fiction. It has been objected to his works that they depend perhaps too much on archæology, the historical details so freely introduced not infrequently competing in interest with the human element; but, without doubt, much of the popularity which *Waverley* and its successors won so easily, and kept so well, was due to the faithful and accurate portraiture of the customs and costumes of a former period, which gave them not only the interest of works of fiction, but all the value of less attractive works of history. Scott was an antiquarian first, and by choice; a novelist afterwards, and from force of circumstances.

In all the illustrations drawn from glove-making or allusions to the craft, Scott shows an intimate acquaintance with the subject, with which no fault can be found, giving another proof—if one were needed—that the Wizard of the North owed his success not so much to magic as a very prosaic industry. One of the most spirited passages in the work—the one most pertinent to the present purpose—is where Simon Glover defends his calling from association with cordwainers (shoemakers), when Henry Gow, the armourer, has coupled the two mysteries together on the ground that both provide for members of the body corporate. Simon, as becomes a good tradesman, will not hear of such an association, and upholds, with some temper, the pre-eminence of the occupation by which he had gathered wealth, and from which, as with many Glovers of our own day, he had derived his surname. "Bethink you," says he, "that we employ the hands as pledges of friendship and good faith, and the feet have no such privilege. Brave men fight with their hands, cowards employ their feet in flight. A glove is borne aloft, a shoe is trampled in the mire ; a man greets his friend with his open hand ; he spurns a dog, or one whom he holds as mean as a dog, with his advanced foot. A glove on the point of a spear is a sign and pledge of faith all the wide world over, as a gauntlet flung down is a gage of knightly battle ; while I know of no other emblem belonging to an old shoe, except that some crones will fling them after a man by way of good, in which practice I avow myself to entertain no confidence."

The sturdy Glover, in his defence of his trade, enumerated the best-known and most enduring ties by which gloves have, from time immemorial, symbolized feelings or declared faith. As a pledge of faith and an emblem of confidence, they had at one time, and particularly at fairs, a very important part to play. Fairs, at one time, filled a large place in business economy. Where towns had not been formed, and the hamlets were not large enough to attract resident traders, fairs afforded the inhabitants the only opportunity they had of supplying their needs, until, in due time, the fair came round again. Considerable business was transacted at these gatherings, which not only brought luxuries and gauds, but carried off the surplus produce ; and it became a matter of business policy to attract as much custom to the fair as possible, and particularly so to the superiors of the neighbouring monastery or church, whose welfare was largely due to the fees or tolls charged, on showing or bringing goods. The fair, in the first place, originated in the congregation of devout worshippers, on the festal day of the saint to whom some church was dedicated. From occasional business being done between people who rarely met at any other time, it became usual to frequent the church festivals for the sake of meeting customers ; the church authorities exacted payment for the privilege, and often gave over the churchyard or some part of the precincts to the traders. After this practice had been unavailingly forbidden by proclamation, an Act of 13 Edw. II. was passed, for the purpose of preventing any further holding of fairs on sacred ground. "The Abbot of Ely, in King John's reign, preached against the holding of fairs on a Sunday. In earlier times, special precautions had been taken to enforce order on sacred ground ; and it was not unusual, when a fair was held within cathedral precincts, to oblige every man to bind himself by an oath at the gate not to lie, steal, or cheat till he went out again " (BLOUNT).

Although the distasteful conjunction between religion and business was severed, the monks did not loose hold upon the tolls of Bord-halfpenny or pitching-pence—fees for ground rent or stall room which were merged in and charged as Hallage, when the fair belonged to the lord of a manor, and the goods were shown in the common hall of a place. Many measures were taken to attract traders. In some instances, freedom from the exactions of barons, who claimed toll on all wares carried through their estates, was granted to all goods brought in to be sold at celebrated fairs. Any established fair had a specified area allotted to it, within which none other could be held; and not infrequently, when it was held near a town, the regular merchants were compelled to close their shops during its continuance. All business was to be confined to the fair; no trafficking or chaffering was allowed on the road; any commodities changing hands before or after a fair were forfeited, and either party to the transaction liable to imprisonment for a misdemeanour. The wealthy citizens of London could claim freedom from toll at any fair they might attend. Ordinary law had no jurisdiction in fairs; when "the peace" of a fair was proclaimed at its opening, perfect liberty was allowed to all and sundry, with immunity from arrest, so that even outlaws and the fugitive bondman might walk openly in confidence through the crowd which thronged the temporary streets of canvas booths, in which, in the Spicery, the Drapery, the Pottery, the Haberdashery, or the Mercery, the various traders were congregated. Did the master meet the runaway thrall, he might neither "chace nor take him" (*Leges Burgorum*). All offences against *the Peace of the Fair*, all disputes as to bargains or the worth of the wares, were referred to a court attached by right to every fair, called, from the dusty feet of the suitors who made plaint therein, the Pie-poudre court. From decisions given therein there was no appeal.

Fairs are likely, before very long, to be wholly discontinued, and to lose even the small vitality which they have yet retained. The cheese fair at Chester, bridge fair at Peterborough, Weyhill fair, horse fairs at Horncastle, Woodbridge, and Howden, and goose fair at Nottingham, still have some pretensions beyond that of being resorts of pleasure-seekers, but the time allotted to them has gradually, but surely, been curtailed. The change since the introduction of railways has been very rapid. It is not so very long ago since Stourbridge Fair—of which the glories once outdid Leipsic, Nuremburg, Frankfort, or Beaucaire, so that it was esteemed the greatest fair in Europe—attracted immense crowds of people from all parts, creating a regular and frequent traffic to and from London, and constituting an occasion on which very considerable business was transacted. In the Duddery, a large open square around which the chief merchants had their stalls, woollen goods to the extent of £100,000 are stated by De Foe to have been sold within a week, and one dealer in light worsted fabrics of Norwich manufacture had a stock there of the value of £20,000. Merchants attended the fair to meet their country customers, to receive accounts, and to take orders.

This is not a digression. It was part of the royal prerogative to set up markets, and fairs were established by virtue of the king's glove, which was the authority under which any free mart or market was held. Thus, says the *Speculum Saxonicum* (*lib. ii.*), " No one is allowed to set up a market or a mint, without the consent of the ordinary or judge of that place ; the king, also, ought to send a glove as a sign of his consent to the same." The glove was ordinarily displayed as a token of security under which trade might be carried on uninterrupted, and was emblematic of the power to maintain order of the king who sent it.

In *Timon of Athens*, the senators ask a glove from Alcibiades before their submission—

> Or any token of thine honour else,
> That thou wilt use the wars as thy redress,
> And not as our confusion ;

and in pledge of protection to all but those who were enemies to the common weal Alcibiades gives his glove. So the glove was borne aloft at a fair in sign of security, a material guarantee of justice and good governance to all the busy concourse of people who flocked thither to chaffer and bargain for the necessaries of life. During the annual fair at Portsmouth, locally known as the " Free Mart," a gilded glove was displayed above the entrance to the White House or gaol, in the High Street. The fair at Southampton, held on Trinity Monday and two following days in each year, was " opened by the Mayor erecting a pole with a large glove to it, and the bailiff then takes possession of the fair. On the Wednesday, at noon, the Mayor dissolves the fair by taking down the pole and glove ; which, at one time, was done by the young men of the town, who fired at it, with single balls, till it was destroyed or they were tired of the sport." (ENGLEFIELD : *Walk through Southampton.*) Correspondents of *Notes and Queries* have offered evidence of a similar custom being observed at Chester, at Newport, in the Isle of Wight, and at Macclesfield ; at Exeter, where the glove was " brought in " with much ceremony, at the commencement of the Lammas Fair, and placed over the Guidhall ; and at Barnstaple, where, previous to the proclamation of the Peace of the fair, a large glove, decked with dahlias, was suspended from a pole protruded from the Quay Hall—the most ancient building in the town—and so remained while the fair lasted. Gorr's *Liverpool Directory*, noting that the town's fairs were held annually on the 25th of July and 11th of November, continues, " Ten days before and ten days after each fair-day, a hand was exhibited in front of the Town Hall, which denoted protection ; during which time, no person coming to or going from the town, on business connected with the fair, can be arrested for debt within its liberty." We may fairly argue that these are but isolated instances of an observance once universal ; and it is probable that a wider search might result in finding many more remains within recent times of this ancient custom.

The glove was then, truly, as Jonathan Oldbuck expressed it, "a sign of irrefragable faith." Truth and trust were so exemplified in gloves that they came to be sworn upon, as though they were relics or holy things. Witness Slender's asseveration of Pistol's guilt, in *The Merry Wives of Windsor*, after accusing him of pocket-picking :—

Ay, by these gloves, did he (or I would I might never come in mine own great chamber again else), of seven groats in mill-sixpences, and two Edward shovel-boards, that cost me two shilling and two pence a-piece of Yead Miller, by these gloves ;

and again, after Pistol has denied the soft impeachment, Slender, with the same oath, re-affirms his guilt—

By these gloves, then, 'twas he.

Biron, too, in *Love's Labour Lost*, binds himself to plain

dealing and plain speech by an oath on "this white glove." Apart from these instances, we know that it was common to swear by the glove, and it is no wonder that, in an age of fantastic profanity, there should have been invocations to "Venus' gloves" and "Mars, his gauntlet" (*Troilus and Cressida*).

According to Hull, gloves were sometimes so recognized as the emblems of trust that they were sent as pledges of safe-conduct in times of truce. He notes it, as the one (and the only one) historical fact connected with the glove that has left a stain upon it, that on the occasion of the marriage of the King of Navarre, the Queen Dowager of Navarre was persuaded to come to Paris by the embassage of a pair of gloves; and then, on the morning of the ceremony, was done to death by poisoned gloves sent to her by René, the court perfumer—her assassination being the prelude to the infamous massacre of St. Bartholomew. The agency of the gloves in this transaction is not now generally credited, although it was generally believed in at the time, and was set forth as indubitable truth by the *Chronicle of France* and other histories. Poisoning by rings, necklaces, and clothing was a common method of wreaking vengeance; and gloves are, in another instance, suspected to have been poisoned to secure the death of an obnoxious person; for Conan, Duke of Brittany, is said to have died, in 1066, from wearing poisoned gloves, at the instigation, it is suspected, of William the Conqueror.

There can be little doubt that the symbolism of security attaching to the glove in this, as in many other associations, arose from its being the covering of the most active and potent member of the body. The strong right hand won and maintained power; it confirmed agreements, and on the top of the sceptre of a monarch, denoted an authority able to reward or punish. It was the hand of honour, and the right-hand glove would appear to have been usually employed in covenants of all kinds. One of Du Cange's citations specially mentions the use of a left-hand glove investiture—indicating that such an instance was exceptional. Thus, the glove represented the hand it usually covered. *They are hand and glove*, says an old proverb, when an unusually close intimacy is to be denoted (FULLER : *Gnomologia*).

Tenures held by gloves are common enough, so much so that Blount says, in the preface to his *Jocular Tenures*, "I have purposely omitted, or but rarely mentioned, those more common tenures, whereby the owner was obliged to deliver, yearly, into the Exchequer, a mew'd Spar-Hawk, a pair of Spurs, Gloves, or the like ; of which kind I met with many, and held them not for my purpose, which was to take in none but what were in some respect or other remarkable." This—so far as this present work is concerned—ominous announcement, is happily neutralized by the citation of three cases in which gloves were the outward and binding sign of a covenant arranged and agreed upon. These are those of "William Drury, who died 7 May (31 *Eliz.*), 1589, and held the manor of Little Holland, in the County of Essex, of the Queen, as of her manor of Wickes, alias Parke-hall, late parcel of the Duchy of Lancaster, by the service of one Knight's fee, and the rent of one pair of gloves turned up with Hare's skin ; " of " John Besett, who (amongst other things) gave to the King 8d. for his relief for 48 acres of Land in Elmesall, Co. York, which John his Father held of the King, by the service of paying, at the Castle of Pontefract, one pair of

Gloves furred with Fox's skin, or eightpence yearly ;" and
of " Phillip Bassett, who held of the King, *in Capite*, the
manor of Wocking, in the County of Surrey, by the service
of half a Knight's fee, and the annual payment of one pair
of Gloves, furred with Grise, to be paid yearly at the King's
Exchequer." The manor of Elston, in Nottinghamshire,
was held by the rent of 1 lb. of cummin seed, a steel
needle, and two pair of gloves.

These, and other like examples, are considered to have
been remains of the ancient practice of binding a bargain,
or transfer of property, by the delivery of a glove ; but, as
regards tenure, might perhaps have begun with the conditions of feudal service, under which lands were held, when
the glove would again be representative of the faith under
which the feofee was bound to do true and laudable service
whenever called upon to fight on behalf of his lord. With
many requirements attached to the holding of land, which
were either demanded by the physical needs of the lord—
such as providing table luxuries at certain seasons, or
doing stated domestic or household service—and more
which were dictated by a spirit of buffoonery, often, with
the coarse humour of our forefathers, becoming flagrantly
indecent or immoral, there was a general symbolism in the
ancient servage of tenants. This can be traced in the
transferred horns and daggers and swords, which were
such common charters of transfers, or gifts of land.
Doubtless, in many instances in which the holder of land
did yearly service in a manner that necessarily degraded
him, there was still probably an undercurrent of purpose
in causing him so to make submission. Grants and gifts
were figuratively made to religious houses ; the dower of
land to a monastery was made by laying a sod of the

given soil upon the altar. Knights figuratively offered
their services to the Church ; it was part of the religious
ceremony in making a knight for the candidate to offer his
sword upon the altar, in token of his devotion to the
interests of the Church. The glove was sometimes, and at
very early times, also made the pledge of a promise. In
offering a gift of lands or other tangible benefits to Mother
Church, a glove was tendered and placed on the altar, as a
sign of fixed purpose ; and we may be sure the good
fathers furthered and upheld the binding character of the
earnest of better things to come. In 1083, the Earl of
Arundel and Shrewsbury vowed the construction of an
Abbey to St. Peter at Shrewsbury, and, in token of his
intent, placed his glove on the altar of the monastery there
(DUGDALE : *Monasticon*).

The glove, from whatever circumstances or connection it
may have derived its significance, entered largely into
transactions in tenure. In the *Rot. Patent* (33 *Hy. VIII.*),
the custom of holding lands by the tenure of a glove
is alluded to as an ancient practice, and the site of the
priory and manor of Worksop, or Workensop, as it was
once known, was then presented to the Earl of Shrewsbury, " to be held *in capite* by the service of the tenth part
of a Knight's fee, and by the royal service of finding a
right-hand glove at his coronation, and to support his right
arm on that day, so long as he might hold the sceptre,
paying, moreover, yearly the sum of £23 6s. 8d." This
service had formerly been attached to the manor of Farn-
ham Royal, in the county of Buckingham, and was due
from the Lords Furnivall till that manor was exchanged
for the Worksop fee by the Earl of Shrewsbury, from
whom it subsequently descended, by inheritance, to the

Duke of Norfolk. This is but one of many other offices of
" grand serjeanty " by which lands are held by doing service at coronations in different capacities.

Bishoprics, and other church offices, were granted by the
delivery of a glove. The Bishops of Paderborn and Mon-
cerce were, in the year 1002, put into possession of their
sees by receiving each of them a glove. In the tenth year
of Henry III., William Briewerr came into the King's
Court, before the king and his great men, and there granted
to Joscelin, Bishop of Bath and Wells, the advowson of the
Church Melverton, in Frank-almoign (that is, free of
secular service), and by his gloves gave the Bishop seisin
of the said advowson (MADOX : *Hist. Exchequer*).

Gloves commonly formed part of mediæval rents, still
probably as retaining the service which was at one time
the first condition of tenure. Instances of such payments,
in kind, are frequently met with, and particularly in the
thirteenth century. Several are given in *Notes from the
Muniments of Magdalen College, Oxford, from the 12th to
the 17th centuries* (W. D. MACRAY).

> 1240. At *Oddington*, Oxon, a pair worth a penny.
>
> 1241. *Shipton-on-Cherwell*, one pair of white gloves or
>   one halfpenny.
>
> 1250. *Swaby*, Linc, and *Brachley*, North Hants, a pair of
>   gloves worth ½d.
>
> 1250-60. At *Guton*, Norfolk, a pair, 1d.
>
> 1260. *Westcote*, two pairs of white gloves.
>
>   *Selborne*, a pair of white gloves worth one penny.
>
> 1290. At *K. Somborne*, Hants, a pair, 2d.

In the reign of Henry III., Walter Granfield devised a
messuage in " the village of Clive," to Henry, son of Henry
de Weever ; paying to Henry, son of Stephen, " chief Lord

of Clive, sixpence a year, reserving to himself a pair of white gloves ;" and, in 1258, Alexander Noke surrendered to the prior of Holy Trinity, Dublin, the lot of ground that he held from him in Gilmaholmog Street, near the church of St. Michael's, for which the prior engaged to pay him yearly two shillings and a pair of gloves, or, in lieu thereof, one penny. It was, says Mr. Leadam, at a later date, when parchment conveyances had superseded all contractual symbolisms, that the transfer of gloves was converted into a payment of glove-money by a purchaser to the steward of the manor ; as an ancient form adds, after fixing the price of the land to be paid to the lord : "Avec les gants de son sergeant estimetz, 20 sols."

As these were symbols of investiture, so the deprival of gloves was made part of the ceremony of degradation. When knights had committed any crime that was capital, " they were stripped of their ornaments, had their military belt took from them, had their spurs cut off with a hatchet, their glove took away, and their arms inverted ; just as it is in degrading those who have listed themselves in the spiritual warfare—the ecclesiastical ornaments, the book, chalice, and such like, are taken from them " (CAMDEN : *The Degrees of England*). This was exactly the ceremony observed in the case of the Earl of Carlisle, in the reign of Edward I., who, after sentence as a traitor had been pronounced upon him, for correspondence with the Scots, was first degraded before his execution (WALSINGHAM).

So, too, as possession was given by granting a glove, they were made tokens of remuneration, another ceremony that was almost wholly symbolical. Bankrupts, in olden time, gave up to their creditors their remaining effects by voluntarily loosing from the waist the girdle, to which was attached their keys and purse, the visible tokens of their possessions. Thus the widow of Philip I., Duke of Burgundy, renounced her right of succession by placing her girdle on the tomb of her husband, even as in renouncing the world before entering on the conventual life ornaments and jewels were taken off during the preliminary ceremony. Du Cange quotes from a charter of the thirteenth century an instance of restitution or re-investiture by the person depositing his glove on the earth.

CHAPTER II

Gloves as Gages.

THE employment of gloves as pledges of personal honour has ceased. The custom, and the combat to which it was the prelude, have been solemnly disfranchised by an Act of Parliament in that case made and provided. But the practice was actually followed within so recent a period, the occasions on which it came to be used were so momentous and full of romantic interest, that, as a picturesque and forcible addition to the language of antagonism, the throwing of the glove in challenge yet lives as a lively and often very annoying simile. In the political differences which constantly arise between parties whose existence is one long disagreement, any challenge to produce documents (which are always being kept back), or to institute enquiries (which are always being evaded), is enforced by an allusion to the old and, it must be admitted, barbarous custom. Particularly where one party professes to court an enquiry, and then, upon being pressed, do not find that the business of the House, or the length of the session, will admit of it being held, the righteous indignation of their opponents rises high, and there is frequent mention of casting down the gauntlet, and the acceptance of the challenge of " my right honourable friend," the enemy.

If the gauntlet were actually cast down on the floor of the House by the indignant seeker after truth, and there yet remained in the present representatives of ancient knights any of the spirit of chivalry that made this the token of knightly truth, there would not.longer remain any cause of complaint, there would be no more evasion or neglect. No challenge enforced by this gage was ever disregarded, unless he who was challenged would rather give up his fair fame and his title to all respect. We cannot picture the gauntlet thrown among our legislators, nor think without a smile of staid statesmen adventuring the justice of their cause and their integrity of purpose on an appeal to arms. The custom is now only fit for caricature, and its memory remains only as a figure of speech. How it appealed once to every sentiment of honour we know from the old tale of the lady who, before the Court of Francis the Fair, threw her glove into the arena of wild beasts, whose conflict all had assembled to witness. The lady, whose preference was so dangerous a boon, took this method of testing the protestations of her lover, and, perhaps, to show her supremacy over a knight of undoubted courage. The incident has more than once been made the text of stirring poems, of which Robert Browning's is, without question, the best.

> Over the rails a glove fluttered,
> Fell close to the lion, and rested :
> The dame 'twas, who flung it and jested
> With life so, De Lorge had been wooing
> For months past ; he sat there pursuing
> His suit, weighing out with nonchalance
> Fine speeches, like gold from a balance.
>
> Sound the trumpet—no true knight's a tarrier
> De Lorge made one leap at the barrier,

Walked straight to the glove, while the lion
Ne'er moved, kept his far-reaching eye on
The palm-tree-edged desert spring's sapphire
And the musky oiled skin of the Kafir—
Picked it up, and as calmly retreated,
Leaped back where the lady was seated,
And full in the face of its owner,
Flung the glove.
              " Your heart's queen—you dethrone her ?
So should I," cried the King ; " 'twas mere vanity,
Not love set that task to humanity ! "

The poet uses his proverbial license to pursue the fortunes of the actors in this little comedy—which might so easily have become tragedy—into later life, when the brave De Lorge proved more submissive and less heroic.

It was in the " wager of battel " that the glove had most prominent employment. This custom is of high antiquity, but the deputation of the glove upon errands of hostility is, most likely, even of more venerable usage, the custom being adopted in legal procedure, because it was fully recognized and respected in private life. In Du Cange, the first example is given in 1499, to which particular attention is called, as being of weighty import : " He gave his promise," the citation runs, " and in token of that, his right hand and the glove of the same." Matthew Paris mentions the incitement to a duel by throwing the glove, under the year 1245. He calls it *Mos Franconum*. The custom was introduced here by the Conqueror, and incorporated among our laws, but its origin is attributed to the Burgundi, a clan of old Gaul, among whose code of laws it is said to be found early in the sixth century. It appears to have been accepted generally by the northern nations and would certainly appeal strongly to their rude hardihood and courage. The wager of battel had three separate jurisdictions—in the court of chivalry and honour, in appeals of felony, and in civil disputes as to possession of real property. The first usage, which was fought to the death, lasted until the seventeenth century; a trial by battel in the court of chivalry, says Blackstone, took place in 1631, and another in the county palatine of Durham, in 1638. The last appeal to force in a civil cause occurred in 1571, and is reported by Sir Henry Spelman, who was himself a witness of the action, which arose out of a dispute as to the ownership of some lands in Kent. The plaintiff appeared in court and demanded single combat. One of them threw down his glove, which was immediately taken up by his opponent on the point of a sword, and the day of combat was appointed. The affair was, however, settled by the Queen's judicious interference. In 1631, Lord Rea impeached Mr. David Ramsey of treason, and offered battle in proof, but the duel was likewise prohibited by King James. Previous to the reign of Henry II., this method of adjusting little differences of opinion was obligatory; it was then made optional by the establishment of the grand assize—a trial before a jury, and being regarded by princes with disfavour, soon became a rusty old legal weapon only dragged out of its scabbard by unscrupulous suitors.

The combats which decided these appeals to Heaven to determine by victory the truth of the parties—for such these trials actually were—do not seem to have been generally very formidable affairs. In the court of chivalry and appeals of felony substitutes were not allowed; the parties to a cause fought it out, although exemption could be claimed in case either of the parties suffered from physical disabilities, or were stricken in years, or of tender age. One of the privileges of the citizens of London was that no wager of battel could hold good against them. Priests could likewise plead their office ; but in civil causes the parties engaged champions as they would lawyers in these days. Churchmen sometimes kept champions in their pay and continual service. Bishop Cantalupe so engaged Thomas de Bruges, at an annual salary of 6s. 8d. The champions, appearing in court, defied each other by severally casting and taking up their gauntlets, and were then taken into custody until the day of appointed trial, which took place before the judges of the Court of Common Pleas and the serjeants-at-law, in lists sixty feet square. The plaintiff and defendant were not allowed to engage each other, lest death should ensue, in which case the suit would have to be stopped in consequence, and no verdict could be given to settle the matter. This appears rather superfluous caution. On the morning of trial the champions were introduced at sunrise into the lists. They had, perhaps, previously attended divine service, and heard one of the liturgies appointed anciently by the Church for duellists, the judge bidding them go to such a church and pray. Then, with bare head and arms and legs, but otherwise well armed, and bearing a four-cornered leather shield and a baton or club an ell long, the champions, each holding the other by the hand, took in turn an oath that the cause they fought for was just and right, and then another oath was made by them both, in this or a similar form : " Hear this, ye justices, that I have this day neither eat, drank, nor have upon me neither bone, stone, nor grass, nor any enchantment, sorcery, or witchcraft, whereby the law of God may be abased, or the law of the devil exalted, so help me God and His saints."

Thus fully armed against ghostly enemies, and tolerably well protected from the blows of his antagonist, each champion commenced to rail at the other, until, in indignation, they fell to blows. Then they pummelled one another all the day long, unless one of them yielded and pronounced the horrible word of craven, or by some unusual mischance happened to get hurt. If sunset came, and the stars appeared, without either gaining the victory, the champion of the actual tenant of the land was adjudged the victor. Possession was even then nine points of the law, and he won by simply maintaining his ground. The punishment of a vanquished champion was various : if he was the champion of a woman for a capital offence, she was burnt and he was hanged ; if he upheld the cause of a man, and that not for a capital crime, the unfortunate champion had his right hand cut off, and his client was closely confined in prison.

The form of battle upon criminal appeals was very similar ; the preliminaries of throwing and taking up the glove were the same, but the oaths taken by the parties were much more solemn ; and hanging of the defendant followed immediately upon defeat, degradation to villainage being the lot of the accuser if he failed to make good his cause. In the court the antagonists took solemn oaths that their cause was just. One, taking the book in his right hand, and in his left the right hand of his adversary, would swear, " Hear this, man, whom I hold by the hand, who callest thyself John by the name of baptism, that I, who call myself Thomas by the name of baptism, did not feloniously murder thy father, William by name, nor am any way guilty of the said felony, so help me God and the Saints ; and this I will defend against thee by my body, as this court shall award." If the plaintiff were defeated he lost all his civil rights, and might be sued for damages. Other ceremonies particular to these occasions are recorded in a passage in Booth's *Nature and Practice of Real Actions,* relative to a dispute occurring in the first year of Henry VI. : "In a writ of right for the manor of Copenhaw, in the county of Northumberland, battle was formed upon the mere right, and the champions appeared. And it was commanded by the court that the champion of the tenant should put five pennies into his glove, in every finger-stall a penny, and deliver it into court, and so the demandants should do the same, and the judges receive the gloves. The champions being on their knees, the council for the parties were asked by the Lord Chief Justice, why they should not allow the champions, and why they should not wage battle, who answered they knew no cause why the duel should not proceed."

On one of the miscellaneous rolls in the Tower, of the time of Henry III., occurs a pictorial representation of one of these legal duels, "rude, it is true, but curiously confirming the testimony that has come down to us of the arms and apparel of the champions" (HEWITT : *Ancient Arms and Armour*). The combatants were Walter Blowberne and Haman le Stare, the latter figuring a second time, as undergoing the punishment attaching to the defeat he experienced, by being hung on gallows, a fate for which his Christian name should have previously prepared him. Both the champions bear quadrangular-bowed shields, and are armed with batons, which are not, however, simple clubs, but topped with a double beak. "An example, agreeing with this description, with the exception of the square shield, appearing to be flat instead of bowed, occurs on a

P

tile-pavement found, in 1856, within the precincts of Chertsey Abbey, Surrey (TIMBS : *Romance of London*).

The last appeal to wager of battel was made in 1818, before the Court of King's Bench, in a notorious trial for murder (*Celebrated Trials, vi.* 227). The body of a girl, named Mary Ashford, was found drowned, with such apparent marks of ill-usage as led to the conclusion that she had met her death by foul play ; and one Abraham Thornton, who was the last person seen in her company, was arrested on suspicion. The cumbrous legal pleadings state that Abraham Thornton was attached to answer William Ashford, who was the eldest brother and heir of Mary Ashford, deceased, of the death of the said Mary Ashford, by choking, suffocating, and drowning. "And the said William Ashford, who was eldest brother, and is heir of the said Mary Ashford, deceased, is ready to prove the said murder and felony against him, the said Abraham Thornton, according as the court shall direct, and hath found pledges to prosecute his appeal." This was upon a verdict of Not Guilty being given by the jury, after a protracted trial, in which the material evidence was very conflicting. The judge, Mr. Justice Holroyd, was satisfied with the verdict, but the relatives of the victim preferred an appeal. The prisoner's counsel seems to have been taken by surprise, and an adjournment took place, when, after a legal argument, the judges allowed the appeal; and the challenge was formally given by throwing down a glove upon the floor of the court by the prisoner, with the words, after he had been called upon to plead, " Not guilty, and I am ready to defend the same by my body." The battle, however, did not take place. Thornton was a stalwart man, and the murdered girl's brother a mere stripling, so that it was considered advisable to abandon the appeal. Thornton is said to have subsequently made confession of the crime. The scoundrel afterwards attempted to reach America, but the sailors, recognizing him, refused to put to sea with such a character on board; and it was only under close disguise that he managed on another occasion to get clear of the country.

Probably this case arose out of a similar appeal then being prosecuted in Ireland. If not, the concurrence of the cases was a remarkable coincidence. Thomas Clancy was tried at the Westmeath Assizes for the murder of Bryan O'Reilly, and acquitted, when the brother of the murdered man sued the court for a writ of appeal of death. The defendant then, on again being arraigned, pleaded that he was "not guilty, and would prove the same by his body." To this "wager" the appellant entered a counter-plea, to which the defendant "demurred." The court, anxious to avoid the combat which should have followed, postponed consideration of the case from time to time, so that two of these appeals to the justice of God, after man had refused it, were together pending. This caused a writer of the time to remark, " Should the duel take place, it will indeed be a singular sight to behold the present learned and venerable judges of the Court of King's Bench, clothed in their full costume, sitting all day long in the open air in Tothill Fields, as the umpires of a match at single-stick. Nor will a less surprising spectacle be furnished by the learned persons who are to appear as the counsel of the combatants, and who, as soon as the ring is formed, will have to accompany their clients within the lists, and to stand like so many seconds and bottle-holders beside a pair of bare-legged, bare-armed, and bare-headed cudgellists."

These anticipations were not realized, and the spectacle, in which many persons would have had great delight, was postponed *sine die.* The Attorney-General brought in a bill "to abolish appeals of treason, murder, and felony," which became law by the 59 *George III., c.* 46, and allowed the release of Thomas Clancy from his imprisonment, and relieved his judges from a somewhat serious predicament, in which, possibly, they saw no occasion of humour.

The casting of the glove in challenge had, as is well known, part and lot in the coronation ceremony. Some people have an idea that this custom is still enforced; but the last appearance of the crown champion at a coronation banquet was when George IV. became king. The head of the Dymoke family being at that time in holy orders, the Committee of Privileges allowed his son to undertake the office. Sir Walter Scott, who was present on the occasion, wrote thus of the champion :—

"The Champion's duty was performed, as of right, by young Dymoke, a fine-looking youth, but bearing perhaps a little too much the appearance of a maiden knight to be the challenger of the world in the King's behalf. He threw down his gauntlet, however, with becoming manhood, and showed as much horsemanship as the crowd of knights and squires around him would permit to be exhibited. His armour was in good taste, but his shield was out of all propriety, being a round *rondache* or Highland target, a defensive weapon which it would be impossible to use on horseback, instead of being a three-cornered leather shield, which in the time of tilt was suspended round the neck. Pardon this antiquarian scruple, which you may believe occurred to few but myself. On the whole, this striking part of the exhibition somewhat disappointed me, for I would have had the Champion less embarrassed by his assistants, and at liberty to put his horse on the *grand pas.* And yet the young Lord of Scrivelsby looked and behaved extremely well. Haydon, the celebrated painter, was greatly pleased with the sight, and deemed it the finest of the day."

The great novelist has not forgotten, in *Redgauntlet,* to make capital of a Jacobite tradition, that the Young Pretender was present at the coronation of George III., and that one of his adherents took up the champion's gauntlet, leaving in its stead his own glove, with a written acceptance of the challenge enclosed in it.   One of the magazines of the day also says that a lady in the gallery dropped, by accident or design, her silk glove, which the herald took up, and in jest handed to the champion, who asked, " Who is my fair opponent ? "   Neither of these incidents appear to have had any foundation in fact, and the latter was particularly contradicted by the herald ; but it seems very probable that the Young Chevalier actually witnessed the instalment of " the Elector of Hanover " upon the throne which he believed to be his by right.   The rumour is mentioned in the *Gentleman's Magazine* for 1764.   David Hume, in a letter written in 1773, records a conversation with the Lord Marshal, who declared his belief, on the authority of one who averred that he had actually seen and spoken to the representative of the Stuarts, that the rumour very generally current only bruited abroad an actual fact.

Between the first and second course of the coronation dinner was the appointed time for the entry of the champion, which was made in order and state as follows :—

Two Trumpets, with the Champion's Arms on their banners.

The Serjeant Trumpeter with his Mace.

The Champion's two Esquires, richly habited, the one carrying
his lance erect upon the right hand, and the
other his shield, with his arms depicted
thereon, on his left.

A Herald of Arms, in his Tabard and Collar, holding
a paper containing the words of the challenge.

| The Earl Marshal on horseback, in his robes and coronet, holding his Marshal's staff. | The CHAMPION, completely armed in white armour, and mounted on a noble horse, holding a gauntlet in his right hand, and having his helmet on his head, ornamented with a plume of feathers of red, white, and blue. | The Lord High Constable on horseback, in his robes and coronet, holding his constable's staff. |

In the *Glory of Regality*, Taylor shows the incidence of the challenge. "The duty of the champion," he says, "is to ride into the hall where the feast of coronation is held, during dinner (before the second course is brought in), mounted on one of the King's coursers, and clad in one of the King's best suits of armour ; he is attended by the lord high constable, and the earl marshal, and by the mouth of a herald is to proclaim a challenge to any who shall deny that the King is lawful sovereign, which, being done, the King drinks to him from a gold cup, which, with its cover, he receives as his fee, and also the horse, saddle, suit of armour, and furniture thereto belonging." The text of the challenge, as used in accordance with set form and recognized usage on the last occasion of its employment was, " If any person, of what degree soever, high or low, shall deny or gainsay our Sovereign Lord, King George the Fourth of the United Kingdom of Great Britain and Ireland, Defender of the Faith, son and next heir to our Sovereign Lord, King George the Third, the last King deceased, to be right heir to the Imperial Crown of this United Kingdom, or that he ought not to enjoy the same, here is his Champion, who saith that he lieth, and is a false traitor ; being ready in person to combat with him, and in this quarrel will adventure his life against him on what day soever he shall be appointed." This was said three several times by the herald, followed in each instance by the casting of the gauntlet, which, when not taken up by any false traitor, was returned by the herald to the champion. The third repetition was at the foot of the throne, and, after an uneventful interval, when a real opponent would have undoubtedly caused no small stir, the champion was pledged by the king in a goblet, which, according to a French account of the ritual of the occasion, was only to be "half-full."

The office has always been honorary, never resulting in an actual encounter, although opposition was anticipated at the coronation of Henry IV. Attached to it, " in grand serjeanty," was the manor of Scrivelsby, in Lincolnshire, to be held " by barony," so that lands and title came with the dignity to the house of Dymoke, whose punning motto is *Pro rege dimico*—I fight for the King. Dugdale and Sir Bernard Burke agree in stating that the grant and appointment were made by the Conqueror to Robert de Marmion, who was made hereditary Grand Champion of England. " For," says Camden, " upon every Coronation of a new King of England, the heir of this family was bound to ride arm'd in compleat harness into the King's hall, and, in a set form, challenge any man to duel, that would dare to withstand the King's right. And this is certain, from the Publick Records, that Alexander Frevill, in the reign of Edward 3, held this same castle (Tamworth) by that kind of service. Yet the Frevills lost this honour in the coronation of Richard 2." Tamworth Castle, to be held by knight's service, was included in Robert de Marmion's grant. The Frevills and Dymocks had come into the family estates by marriage with the daughters of a direct descendant, in whom the male line ceased ; but, in the dispute

which arose between them as to who should undertake the
office of champion, the right was adjudged to the Dymocks
on "more anthentick Record and evidences," and from that
period until now (for the office is not abolished) they have
continued to hold and enjoy their high office and its
privileges, taking no account of change of dynasty, but
doubtless ready to perform their stipulated service for
the king appointed, without caring about his descent.

As a symbol of defiance, the gauntlet is thrown upon
the field by Entellus, in the fifth book of the *Æneid*, in
the stirring account of his fight with Dares, the Trojan,
during the funeral games promoted by Æneas. No report
of a battle in the palmy days of the prize ring ever
equalled this for graphic vigour, and many translations of
the encounter (of which we have taken Conington's [1873]
for choice) bring the scene vividly before the mind's eye.
The gauntlets of the combatants are prominent throughout
the fray. Æneas first calls aloud for athletes to come and
contend for the golden-horned bull that should reward the
victor :—

> Stand forth, your wrists with gauntlets bind,
> And lift your arms on high.

Dares at once appears in the lists, but his opponent is
sought and sought in vain : none ventured to measure
their skill with the conqueror of Butes and opponent of
Paris—

> Of all the assembled throng, not one
> Has nerve the champion to defy,
> And round his hands the gauntlets tie.

He, impatient of delay, claims the coveted reward, and
all his Trojan friends, with shouts, uphold his claim. It
appears as if he must be allowed a bloodless victory, but

Acestes chides Entellus, pupil of Eryx, famed throughout Trinacria, with his inaction, and spurs him to try once more his old-accustomed skill. No coward dread has kept him back, Entellus answers, nor would a prize twice as fair tempt him, but he will not let the bull go without a struggle:

> Then on the ground, in open view,
> Two gloves of giant weight he threw,
> Which Eryx once in combat plied,
> And braced him with the tough bull-hide.
> In speechless wonder, all behold
> Seven mighty hides with fold on fold ;
> Enough the first, and iron sewed,
> And knobs of lead augment the load.

These mighty "mufflers," as boxing gloves were once known, are thrown forward before the wearer appears to make good the promise they imply ; and this custom had a curious counterpart in the practice of pugilists, who always cast their hats into the ring first, before entering it themselves. The gauntlets are, in themselves, sufficient to deter Dares from further action ; he relinquishes all thought of antagonism with a man who can wield such deadly weapons. The old hero, once warmed into action, is not to be thwarted. He says to Æneas :—

> " What if the gauntlets you had seen
>   Alcidas wore that day,
> Had stood on this ensanguined green,
>   And watched the fatal fray ?
> , These gloves your brother Eryx wore,
>   Still stained, you see, with brains and gore ;
> With these 'gainst Hercules he stood ;
> With these I fought, while youthful blood
> Supplied with strength, nor age had shed
> Its envious winter on my head.

> But if the arms Sicilians wield,
> Deter the Trojan from the field,
> If so Æneas' thoughts incline,
>   And so my chief approves,
> Let both be equal, side and side :
> I spare you Eryx' grim bull-hide,
> Dismiss that terror, and resign,
>   In turn, your Trojan gloves."

Equally armed with gauntlets matched in size, the combatants face each other, and the fight begins. Lacking agility, and stiff in limb, Entellus stands still, while Dares, first on this side, then on that, feints, and tries now to break down the skilled defence, or to tempt the old athlete to advance. All the while—

> 'Mid ears and temples come and go,
> The wandering gauntlets to and fro.

until, at last, Entellus aims a blow, which is dexterously avoided, and falls prone upon the ground.

The anxious throng raise shouts and answering shouts, but the battle is not over. Entellus is raised—a new man. Wrath spurs him on, and sends his aged blood coursing through his veins with new vigour. He turns again to the fray in a different spirit, and forsaking his defensive plan, he " fights forward." Dares is unable to stand before him, and is in sore danger of being then and there killed outright, but Æneas stops the fight before the awakened thirst of strife carried Entellus beyond control. Dares is led off by his friends, and the bull and bay awarded to the victor.

> With triumph kindling in his eyes,
> And glorying in the bull, his prize,
> The victor to the concourse cries :
> " Learn, goddess-born, and Ilium's host,
> What strength my youthful arm could boast,

And what the death from whose dark door,
Your rescued Dares you restore."
He spoke, and stood before the bull,
Swung back his arm, and planted full
    Between its horns the gauntlet's blow.
The brain came through the shattered skull ;
    Prone, quivering, dead, the beast lies low ;
While words like these the veteran said,
In consecration of the dead :
" This better substitute I pay,
    Eryx, to thee, for Dares' life,
And here renounce, as conqueror may,
    The gauntlets and the strife."

These deadly combats, fought with such cruelty and
determination, were forbidden by the laws of Lycurgus
to the Lacedemonians. The gauntlets used with such
fearful effect are generally held to have been composed of
leathern thongs, studded or stuffed with lead or iron, to
give additional force to the blows, but some critics have
claimed them to have been a kind of club or bludgeon
with lead at one end, or with leaden balls suspended
from it.

Having looked upon this picture from Virgil, let us look
upon one from Shakespeare. In *Henry V.* a glove is
given quite a prominent part in the lighter scenes, which
relieve the weightier issues—the course of crowns and fate
of kings—on which the tragedy turns. Noble Harry,
going through the camp on the eve of Agincourt, enters
into conversation with Williams and Bates, two of his
soldiers. The talk turns on the imminent battle, and upon
the responsibility of rulers for the slaughter in warfare, and
the dispute, in which Henry, without disclosing himself,
defends his own reputation, gets so hot as to come
near a personal quarrel, so that Henry says, "Your

reproof is somewhat too round ; I should be angry with you, if the time were convenient." The other is quite willing to postpone the matter, and so, with a significant proviso, asks, "Let it be a quarrel between us, if you live."

KING HENRY.   I embrace it.

WILLIAMS.   How shall I know thee again?

KING HENRY.   Give me any gage of thine, and I will wear it in my bonnet : then, if ever thou darest acknowledge it, I will make it my quarrel.

WILLIAMS.   Here's my glove : give me another of thine.

KING HENRY.   There.

WILLIAMS.   This will I also wear in my cap : if ever thou come to me and say, after to-morrow, *This is my glove,* by this hand, I will take thee a box on the ear.

KING HENRY.   If I ever live to see it, I will challenge it.

WILLIAMS.   Thou darest as well be hanged.

KING HENRY.   Well, I will do it, though I take thee in the King's company.

WILLIAMS.   Keep thy word : fare thee well.

The battle is fought, with what result it is not necessary to write in any English book, and on the field the King, catching sight of Williams, has him brought, and, confident that the darkness has hidden his identity, asks—

KING HENRY.   Soldier, why wear'st thou that glove in thy cap?

WILLIAMS.   An 't please your Majesty, 'tis the gage of one that I should fight withal, if he be alive.

KING HENRY.   An Englishman?

WILLIAMS.   An 't please your Majesty, a rascal, that swaggered with me last night ; who, if a' live, and ever dare to challenge this glove, I have sworn to take him a box o' the ear: or, if I can see my glove in his cap (which he swore, as he was a soldier, he would wear, if alive,) I will strike it out soundly.

The king refers to Fluellen, a choleric Welsh captain, the question of military etiquette, whether the man should redeem his pledge, even though the unknown might be a gentleman of great sort, far removed from Williams in degree. Fluellen has no doubt in the matter, though, " he be as goot a gentleman as the tevil is, as Lucifer and Beelzebub himself," yet Williams should hold to his oath and challenge the glove, upon which Henry dismisses him with the direction, " Then keep thy vow, sirrah, when thou meet'st the fellow." No sooner is he gone, than the king turns to Fluellen, and gives to him the glove received in gage from Williams the night before:—

> Here, Fluellen ; wear thou this favour for me, and stick it in thy cap : when Alençon and myself were down together, I plucked this glove from his helm : if any man challenge this, he is a friend to Alençon, and an enemy to our person ; if thou encounter any such, apprehend him, an thou dost love me.

The honest Welshman is delighted: " Your Grace does me as great honours as can be desired in the hearts of his subjects : I would fain see the man that has but two legs, that shall find himself aggriefed at this glove, that is all ; but, I would fain see it once, and please God of His grace that I might see it."

His desire is gratified much sooner than he anticipated, for the king sends him off on an errand to Gower, the officer in whose company Williams served, and directs the Duke of Gloster and Earl of Warwick to follow him, and see that no harm ensues from the quarrel almost certain to follow. Williams and his captain are conversing together as Fluellen comes up. No sooner is the message delivered, than Williams strikes in—

WILLIAMS. Sir, know you this glove ?
FLUELLEN. Know the glove ! I know the glove is a glove.
WILLIAMS. I know this ; and thus I challenge it,—

giving him, at the same time, with good will, the box on the ear promised to the wearer of the glove. In the outcry that ensues on the sudden and, to those that stand by, unprovoked assault by a soldier on an officer, Fluellen tries to return the blow, calling the while for the offender to be arrested as a traitor to the King and his cause. The nobles come up, and, immediately afterwards, the King, to whom Fluellen turns:—

FLUELLEN. My liege, here is a villain and a traitor, that, look your Grace, has struck the glove which your Majesty is take out of the helmet of Alençon.

WILLIAMS. My liege, this was my glove: here is the fellow of it: and he that I gave it to in change promised to wear it in his cap; I promised to strike him if he did: I met this man with my glove in his cap, and I have been as good as my word.

FLUELLEN. Your Majesty hear now (saving your Majesty's manhood) what an arrant, rascally, beggarly, lousy knave it is: I hope your Majesty is pear me testimony, and witness, and avouchments, that this is the glove of Alençon, that your Majesty is give me, in your conscience, now.

KING HENRY. Give me thy glove, soldier; look, here is the fellow of it. 'Twas I, indeed, thou promised'st to strike; and thou hast given me most bitter terms.

FLUELLEN. An please your Majesty, let his neck answer for it, if there is any martial law in the 'orld.

KING HENRY. How canst thou make me satisfaction?

WILLIAMS. All offences, my liege, come from the heart: never came any from mine, that might offend your Majesty.

KING HENRY. It was ourself thou didst abuse.

WILLIAMS. Your Majesty came not like yourself: you appeared to me but as a common man; witness the night, your garments, your lowliness; and what your Highness suffered under that shape, I beseech you, take it for your own fault, and not mine: for had you been as I took you for, I made no offence; therefore, I beseech your Highness, pardon me.

KING HENRY. Here, uncle of Exeter, fill the glove with crowns,
And give it to this fellow.—Keep it, fellow,
And wear it for an honour in thy cap,
Till I do challenge it.—Give him the crowns:—
And, captain, thou must needs be friends with him.

FLUELLEN. By this day and this light, the fellow has mettle enough in his pelly:—Hold, there is twelve pence for you, and I pray you to serve Got, and keep you out of prawls, and prabbles, and quarrels, and dissensions, and I warrant you, it is the petter for you.

WILLIAMS. I will none of your money.

FLUELLEN. It is with a goot will ; I can tell you, it will serve you to mend your shoes : Come, wherefore should you be so bashful ? your shoes is not so goot ; 'tis a goot silling, I warrant you, or I will change it.

To throw or send the glove has been a mark of defiance from very early times. When it began, or whence it came, we cannot tell, but the custom, in all probability, was not known here until after the . Conqueror came over. An old romance, of the fourteenth century, *Amis and Amiloun*, notices it :—

> Yea, sayd the duke, wilt thou so ?
> Dar'st thou into battle go ?
> Yea, certes, sayd he tho',
> And here my glove give I thereto.

" Panormitanus makes mention of one Duke Reynard, who, by a Herald, sent a gauntlet or glove unto Alphonsus, King of Arragon, and withall denounced him battell ; who willingly accepted the same, and demanded of the Herald whether he challenged him to fight with his army, or in single combat : who answered, Not with his army ; whereupon Alphonsus assigned a day and place for the purpose, and came at the prefixed time, but the Duke failed." Old Guillim, noting this incident in his *Display of Heraldrie,* when treating of the coat of arms of the Earls of Westmoreland, which has three left-hand gauntlets, topaz, on a sapphire field, takes occasion to remark that the law of challenges gives the choice of weapons to the party defendant—a courtesy which made part of the etiquette of duelling in modern times. He calls in witness Sprigellius : *Jure belli licet provocato diem and locum Prelii dicere.* It would appear from the *Vision of Piers Plowman,* a satirical poem, generally attributed to William Langland, a secular priest, who lived about the time of Edward III., that it was used in challenge by others than nobles.

> And then gan a wastoure to wrath him, and wold have fought,
> And to Piers the Plowman he profer'd his glove.

Among the northern borderers the glove, as a gage, held high place. By the glove, actions of their rude life were scrupulously ruled ; it was their law where other law there was none. Did one of them break faith ? The surest remedy was for the injured person to appear at the next common meeting-place, and ride through the assemblage, bearing a glove on the point of a lance, proclaiming the perfidy. The symbol roused so keen a sense of right, so fervently appealed to their rough justice, that the offender was often slain by his own clan to wipe out the disgrace brought upon them. To bite the glove, or the thumb, which, says the serving-man in *Romeo and Juliet,* is "a disgrace to them, if they bear it," was the sign of hostility and sure prelude of a quarrel.

> Stern Rutherford, right little said,
> But bit his glove and shook his head.
>
> —*Lay of the Last Minstrel.*

" It is yet remembered," writes Scott, " that a young gentleman of Teviotdale, on the morning after a hard drinking-bout, observed *that he had bitten his glove.* He instantly demanded of his companion with whom he had quarrelled, and, learning that he had had words with one of the party insisted on instant satisfaction, asserting that, though he remembered nothing of the dispute, yet he was sure he never would have bit his glove unless he had received some unpardonable insult. He fell in the duel, which was fought near Selkirk, in 1721."

In the rare *Life of Bernard Gilpin,* the fearless border apostle, it is recorded that he observed a glove hanging high up in the church to which he was attached, which had been placed there in consequence of a deadly feud prevailing in the district, and which the owner had hung up in defiance, daring anyone to mortal combat who ventured to take it down. Gilpin asked the sexton to remove it. "I dare not," was the reply. The pastor then called for a long staff, took down the emblem of enmity, and placed it in his bosom. In a sermon, soon after, he inveighed particularly against the barbarous custom of challenges. "I hear," he continued, "that there is one among you, who, even in this sacred place, hath hanged up a glove to this purpose, and threatened to enter into combat with whomsoever shall take it down. Behold," producing the glove at the moment, "I have taken it down myself." Through fact to fiction, this recalls the scene in the High Church of St. John, in *The Fair Maid of Perth,* when Bonthron attempts to cover his murder of Oliver Proudfute by offering, in the presence of his dead victim, "the combat to any man who says I harmed that dead body," and, according to usual form, he threw his glove upon the floor of the church. "Henry Smith stepped forward, amidst the murmured applause of his fellow-citizens, which even the royal presence could not entirely suppress, and lifting the ruffian's glove, placed it in his bonnet, laying down his own in the usual form, as a gage of battle." Justice trod hard on crime, and in the combat which followed, the sturdy smith defeated his antagonist, who was, as demanded by

Q

the law of appeal, at once taken to the gallows and executed. The combative significance of the glove was still maintained as late as 1784, when, according to Grose, it was still so employed among the Highlanders and on the Continent.

CHAPTER III.

Gloves as Gifts.

IN direct contrast with the employment of gloves on errands of hostility, we have now to consider them as messengers of goodwill sent to bridge over breaches of friendship, and bind up broken ties, and offered on memorable anniversaries as tokens of amity, of fidelity, and kindly feeling. Once more we must take it for granted that gloves, in these more humane, if not more human, relationships, must be considered as only representing those who gave them ; that they were mute ambassadors, silent deputies, offering in token of unity the covering of the hand, which, with a clasp, would corporally confirm an avowed purpose, or cement an alliance. Men once used osculation in greeting, and proffered, on meeting a friend, the kiss of peace. But the standard of honour was not high, and the salute was far too close—too convenient a cloak for perfidy. In unsettled society man mistrusted his neighbour, and was ever on the lookout for ambuscades and treachery. It suited him better to offer, and take, the right hand of those with whom he had transactions, so that he had, in giving a greeting, not only a personal bond, but what was fully as much to the purpose, a good guarantee against surprise in holding the weapon hand, which would generally be used against him.

Q 2

New Year's Day was the great day for making gifts of gloves. The anniversary is experiencing a considerable revival with us of late years, and the sentimental saint of February is gradually losing ground. Formerly the calends of January were kept with feasting and mirth, with visits of carol-singers bearing bowls of wassail of which all should drink, with mummings and maskings, and exchanging of garments between men and women, and sundry other observances leading to the conclusion that the Sigillaria of the ancients had in some sort been retained and kept up. The license allowed at this time, and to the saturnalia which preceded it, led to many ecclesiastical injunctions and no small restriction of the usual festivities among the faithful. But with the common wish of a good or happy New Year freely expressed on all hands, as well as with the ordinary practice of making presents at this season, there was no interference; it was a simple kindly custom to which none but a churl could object. It was universal. The king on New Year's Day, early in the morning, received with much ceremony a gift from the queen; members of a family presented them to each other; friends and aquaintances in courtesy sent some token of remembrance; retainers and servants made presents to their lords and masters.

The gifts were not always spontaneous, but given and rewarded according to a recognized rule. In the *Household Book of the Earl of Northumberland* are the two following items :—

Item, my Lorde usith and accustomyth to gyfe yerely when his Lordshipe is at home to his iij Hanshmen uppon New-yers-Day, when they doo gyfe his Lordschip glovis to his New-yers-gyft. And in reward after vjs. viijd. to every one of them.

Item, My Lord usith and accustomyth to gyfe yerely upon New-yers-Day to every of his Lordschipes Footmen when they doo gyfe his Lordship gloves for his New-yers-Gyfe the said New-yers-day in the mornnynge. And in reward as his Lordship is accustomede yerely iijˢ· iiijᵈ· to every of them.

These items relate to the custom of rewarding givers with a present in return for their generosity—a custom on which it may be supposed the tendering parties not seldom reckoned on receiving more than they gave. Elizabeth was not, in slang phrase, to be "had" in this manner. She had presents in plenty, particularly, as the good old custom was, on New Year's Day; "but," says Dr. Drake, "though she made returns to her New Year's gifts, she took sufficient care *that the balance should be in her own favour.*" For some time after New Year's Day, gifts were tendered to those who were were late in coming. The manuscript Day Book of the Countess of Pembroke records the present to her cousin's wife of a pair of buckskin gloves on the 10th of January, and another pair to a visitor on the 17th of the month. In another entry, the Countess enters the gift to a Mrs. Winch, of Settra Park, of "four pair of buckskins that came from Kendall."

The presents on these occasions generally took the form of small articles of apparel—hose, bonnets, lace, even elaborate "smocks"—but gloves, as might be expected, were more frequent than any. Instances of these have already occurred in these annals; and it will be sufficient to complete the enumeration of the gloves given to Mary in 1556, in addition to those already quoted :—

By Baker, Confessor, foure paire of gloves, two of them furred, thother two lined.

By Mrs. Zyzans, a peire of gloues wrought with silke.

By Phillip Manwaring, two peire of gloues.

By Frauncis Euerarde, two peire of gloues.

By the Henchmen, a peire of gloues.
By Anthony Lambortye, a peire of gloues.
By Anthony Anthony, a peire of gloues with lowpes of golde, lyned with crymson vellat, in a boxe.
By Zenzan's two sonnes, 2 peire of gloues.

There is a passage in the *Byting Satires* of Bishop Hall, denoting a practice of giving gloves on the occasion of another festival, anciently observed with far more ceremony and rejoicing than now. The censure is of "some base hedge-creeping Collybist," who—

> Scatters his refuse scraps on whom he list,
> For Easter gloves, or for a shrove-tide hen,
> Which, bought to give, he takes to sell again.

The reference is not very flattering to any party to such a transaction, but there is proof in this that gloves were in some sort made Easter dues, and in some measure had part in the relaxation and general joy with which the great Church holy-day was once celebrated.

Special occasions of local importance, the commencement or conclusion of some important labour, were marked and recognized by gifts of gloves. In the *Shuttleworth Accounts* (*Chetham Soc.*), on the occasion of laying the first stone of Gawthorpe Hall, in August, 1600, " 2s. 6d. was given to ten labourers, to buy them every one a pair of gloves; to Anthony Whitehead, the master mason, and five masons, for the same purpose, 2s. 2d.; to Jane Hodgkinson, the housekeeper, and two maids (amount obliterated); to eight male servants, including the cowboy, 21d." This should show a far more general wearing of gloves in former days than now, if the money given was always expended in the manner indicated. First, we find mention of "gift-gloves," afterwards, of "glove-money," or "glove-silver," given to servants as an equivalent for the gloves which long-established usage had led them to expect, but which, as money grew more plentiful, were in many instances probably received in embarrassing quantity. The glove-silver was certainly not the money which "lined" the gloves, when the gift was required to be proportionate in value to the rank of the recipient.

There are other notices of gloves being purchased in this volume, which are, for the information they give as to prices, of some value :—

> March, 1612. A pair of gloves to Turner, 8d.
> November, 1612. A pair of gloves to Abel, 5d.
> June, 1613. A pair to Mr. Barton, 6d.
> November, 1617. Two pair, 2s.
> June, 1620. Three pair to the gentlemen, 6d.
> July, 1620. Given to Thomas Smythe a pair, 2s.
> Three pair, 3s. 6d.

Several of these items apparently relate to gifts to the servants—a practice, even at that time, of considerable antiquity. In Bishop Fleetwood's *Chronicum Preciosum* the following entries appear under the year 1425 :—

> For thirty pair of autumnal gloves for the servants, 4s.
> For twelve pair of gloves to the Bishop of Worcester's servants, 5s.

This leads us to the gift of harvest gloves to labourers after they had finished their hard work of gathering in the fruits of the earth. Old Tusser, in his *Five Hundred Points of Good Husbandrie*, 1580, enjoins—

> Grant harvest lord (an overlooker), more by a penie or twoo,
> To call on his fellowes the better to doo ;
> Give gloues to thy reapers a larges to crie,
> And dailie to loiterers have a good eie.

Largess, Bailey gives as a " free gift, a dole, or present," and Cotgrave as " bounty, handfuls of money cast among the people." In Huloet's *Dictionary* of 1552, it is rendered, " crie a larges, when a rewarde is given to workmen. *Stipem vociferare.*" The quaint verse of Tusser (*English Dialect Society*) is well illustrated by Major Moor's *Suffolk Glossary.** Largess is there explained as " usually a shilling. For this the reapers will ask you if you ' choose to have it hollered.' If answered, Yes, they assemble in a ring, holding each other's hands, and in- clining their heads to the centre. One of them detached a few yards apart, calls loudly, Holla, Lar ! Holla, Lar ! Holla, Lar ! jees. Those in the ring lengthen out, O-o-o-o, with a low, sonorous note and inclined heads, and, then, throwing the head up, vociferate, A-a-a-a-ah ! This, thrice repeated for a shilling, is the established exchange in Suffolk." Can we not well imagine this to be a time-honoured custom since the days of Tusser, and to have been then rewarded by gloves instead of a shilling, which would represent glove-money. In the *Household Books of the Earl of Northumberland* are entries of yearly gifts to officers and servants for crying Larges before his Lord-ship on New Year's Day. Tusser says further—

> In Maie get a weede hooke, a crotch, and a gloue,
> And weed out such weedes as the corn doth not loue ;

leading us to another bit of agricultural lore : Fitzherbert (*Boke of Husbandry*, 1586) enumerates, as " ye chyef in-strumentes for weeding, a paier of tonges made of wood, and in the farther end it is nicked to hold ye wede faster. . . . .

---

* Baker's *Northamptonshire Glossary* also gives Largess as a local term, significant of a harvest gift.

Yf it be drye wether, then must ye have a wedying hoke with a socket set upon a lytle staffe a yard longe. And this hoke wolde be wel steled and grounde sharpe bothe behynde and before. And in his other hande he hath a forked stycke a yarde long."

The farming of former days was anything but scientific ; indeed, not to mince matters, it must have been very haphazard and slovenly. Thistles grew freely with the corn, so that gloves—probably of thick leather, like the dannocks, or hedging gloves, of labourers in our time—were necessary in reaping the corn, and were used, too, to gather the thistles from out the growing crop, wherewith to feed the cattle. Among the expenses of the Priory of Holy Island for 1344-5, is an entry of 2s. 8d. for "Gloves for 14 servants when they gathered the tithe-corn." In Sir J. Cullum's *History of Hawsted* is mention of "five pairs of harvest gloves, 10d.," and a statement that "the Monastery of Bury allowed its servants twopence each for glove-silver in autumn." The rural bridegroom, in Laneham's *Account of the Entertainment of Queen Elizabeth at Kenilworth Castle*, 1575, has "a Payr of Harvest Gloves" on his hands, as a sign of good husbandry.

A curious observance, having kin to this claim of servants to gloves as their due, is recorded by Chambers, who says it is not safe to enter the stables of princes without pulling off the gloves. On this D'Israeli says that it is an ancient established custom in Germany that whoever enters the stables of a prince, or great man, with his gloves on his hands, is obliged to forfeit them, or redeem them by a fee to the servants. The same custom is observed in some places at the death of the stag ; in which case, if the gloves are not taken off, they are redeemed by money given to the huntsmen and keepers. The French king never failed of pulling off one of his gloves on that occasion.

There is that other claim of gloves required in reward by ladies who have the hardihood to steal a kiss from a sleeping man, which Sir Walter has, again, not forgotten to utilize in *The Fair Maid of Perth*, when Catherine, on St. Valentine's morn, ventures to kiss Henry Smith when she finds him asleep, and which Gay mentions in his *Pastoral* :—

> Cicely, brisk maid, steps forth before the rout,
> And kiss'd with smacking lip the snoaring lout ;
> For custom says, " Whoe'er this venture proves,
> For such a kiss demands a pair of gloves."

The folk-lore of gloves was in this respect varied by allowing any person who first saw a new moon through glass to take from the man nearest at the time either a kiss or a pair of gloves. (HONE.)

In this amatory sport, Herrick, the rollicking parson-poet with such unparsonlike tendencies, makes a kiss the reward in a sport called Draw-glove, a game played by holding up the fingers, representing words by their different positions (HALLIWELL). The reward, by a delightful arrangement, would be likely to be equally agreeable either in winning or losing:—

> At Draw-glove we'll play,
> And prithee, let's lay
> A wager, and let it be this ;
> Who first to the summe
> Of twenty doth come,
> Shall have for his winning, a kisse.
>
> —*Hesperides.*

In the same work the parson of Dean Prior makes allusion to the distributing of gloves at weddings :—

> What Posies for our Wedding Rings,
> What Gloves we'll give and Ribanings.

This is a very ancient association of gloves. Dekker makes allusion to "the innocent white wedding gloves." Rare Ben Jonson, in his play of *The Silent Woman*, makes Lady Haughty observe to Morose, "We see no ensignes of a Wedding here, no Character of a Bridall. Where be our Skarves and Gloves?" The Clown in *The Winter's Tale*, playing, alludes to the cost of buying gloves at the ceremony: "If I were not in love with Mopsa, thou shouldst take no money of me; but, being enthralled as I am, it will also be the bondage of certain gloves." So, too, Beaumont and Fletcher, in the *Scornfull Ladie*, not only mention this, but several other minor ceremonies associated with marriage :—

> Believe me, if my wedding smock were on,
> Were the gloves bought and giv'n, the license come,
> Were the rosemary branches dipp'd, and all
> The hippocras and cakes eat and drunk off,
> Were these two arms incompass'd with the hands
> Of bachelors, to lead me to the church,
> Were my feet at the door—were "I John" said,
> If John should boast a favour done by me,
> I would not wed that year.

A poetical description of a trades procession in the eighteenth century says—

> Next March the Glovers, who with nicest care
> Provide white kid for the new-married pair,
> Or nicely stitch the lemon-colour'd glove
> For hand of beau, to go and see his love.

The cost of complying with this social exaction was often very heavy, for the gloves were not only distributed among those actually present on the occasion, but were sent broadcast to all who had any title to be considered friends or acquaintances of either of the contracting parties. For the wedding, in 1567, of the daughter of Mr. More, of Losely, there were purchased—

One dozen of gloves	...	...	...	...	...	10s.
One other dozen of gloves	...	...	...	...	...	5s.
iij dozen of gloves at iijs. a dozen		...	...	...	...	9s.

A letter sent from the "Cort at Whitehall," on the 28th December, 1604, from Sir Thomas Edmonds to the Earl of Shrewsbury, gives an account of the marriage "of Sr. Phillipp Harbet and the Lady Susan, yesterdaie, sollempnised wth great honor. The King and Queen assisting the same in the Chappell. She was led to the Chappell by the Prince and Duke of Holst, and brought back by the Lord Thr'er and the Lord Admyrall, and she geiven by the King: The Cort great in nomber of Lordes and Lades, and both sortes magnificant in braverie: The charge of the Gloues and Garters geiven esteemed to amounte to well neare a thousand poundes: but the same was recompensed in the p'sents of plate wch were geiven to a great vallue." (LODGE: *Illus. of Brit. History.*) Another letter, relating the same incident, from Mr. Winwood to Sir Dudley Carleton, says, "No ceremony was omitted of Bride-cakes, Points, Garters and Gloves."

Not only at weddings, but in the pre-contracts or legal betrothals, which very generally preceded them, gloves had a conspicuous place. On such occasions it was general for the lover to give his mistress either a bent sixpence or a pair of gloves. " One lover, who was betrothed in the same year in which Shakespeare was engaged to Anne Hathaway, gave also a pair of gloves, two oranges, two handkerchiefs, and a girdle of broad red silk." Pieces of bowed money are still believed to bring good luck, and to be gifts of happy omen, and have been given as presents from time immemorial. They were given in evidence of goodwill between relations, as well as between lovers, and round them cluster a number of superstitions. That gloves should be used in these preliminary espousals, which had such legal acceptance that neither party to a pre-contract could be married to any other person, is only what we might expect from all their traditions. In 1571, there came on a trial between Thomas Soley and William Headley, of the one part, and Agnes Smith, of the other part, where each of the men alleged that Agnes Smith had bound herself to him by a formal betrothal. It was proved that the ceremony with Headley had actually been performed, and on the occasion he presented her with two crooked coins, and she, for her part, gave him a gold ring. One witness declared that he had carried from Headley to Agnes " a bowed grote and a bowed sixpence, as tokens." The defendant admitted the pre-contract with Headley, but denied that with Soley. Being engaged to the one, she said, she could not have entered into any matrimonial agreement with the other. Soley offered her a pair of gloves, but she refused them. She plainly considered that ' the acceptance of the gloves would have compromised her.

The clergymen officiating at weddings once shared with the guests in the lavish gifts of gloves. It is now as rare for a clergyman to exact the toll he was once expected to take from the lips of the bride, as it is for him to receive the fee of gloves, which he once seems not only to have looked for as a courtesy, but demanded as his due.

In Arnold's *Chronicle* (about the date of 1521), chiefly concerning London, among "The artycles upon whiche is to inquyre in the Visitacyons of Ordynaryes of Chyrches," we read—"Item, whether the Curat refuse to do the solemnysacyon of lawfull matrymonye before he have gyfte of money, hoses, or gloves."

An Act of Parliament of the 26th year of Henry VIII., leaves no doubt that the priesthood in those days made large and unreasonable demands on these occasions. The levying of Commorth is there prohibited, this being an enforced contribution at weddings, and taken, too, from young priests saying their first mass, making them, as a common phrase still current would put it, "pay their footing." Brand shows a similar demand made in Belgium, where it was the custom for the priest to ask of the bridegroom the ring, and, "if they could be had, a pair of red gloves, with three pieces of silver money in them; then, putting the gloves into the Bridegroom's right hand, and joining it with that of the Bride, the gloves were left, on loosing their right hands, in that of the Bride." (SELDEN: *Uxor. Hebraica Opera*). In the *Popular Antiquities,* too, is related how, at Wrexham, in Flintshire, on occasion of the marriage of the surgeon and apothecary of the place, August, 1785, Dr. Lort saw, at the doors of his own and neighbours' houses, throughout the street where he lived, large boughs and posts of trees, that had been cut down and fixed there, filled with white paper cut in the shape of women's gloves, and of white ribbons.

The mingled yarn of the web of life is here of equal

threads; the same ceremony observed—how generally is not known—by this antiquary is one of the most pathetic and tender associations of that long sleep which rounds our little life. How the same manner used in keeping alive the remembrance of the dead came to be used in marking a marriage with some show of gratulation and rejoicing it would be hard to say. It is certain that crowns of flowers have been used at funerals for ages past, not only upon the heads of mourners, but, as they are now employed, in strewing them upon the coffin, sometimes upon the corpse, of a relative or friend. It would be easy to expend much matter in proving this point, and in showing how the wreath or circlet had, like the ring, peculiar symbolism, and was in many instances particularly consecrated by the ancients to certain deities. From the practice of crowning victorious heroes and winners in athletic contests with chaplets of leaves, they took to giving the dead this sign of conquest, and then kept green the memory of their loved ones by continuing to embellish the graves, which alone remained to constant grief, with similar tokens. Thus the practice, common in our churchyards, of rearing plants and flowers on the graves, no less than that of laying memorial wreaths upon a coffin at a funeral, are distinctly connected with the Olympian games, by which the Greeks marked the progress of time, and in which victory was held not only to have conferred immortality upon the winner, but to have brought to his country an inestimable good. So the circlets of flowers solemnly used in bringing departed friends to their rest in God's Acre, crown the memory of the dead with honour, and give to those who have run their course and fought their fight the prize allotted to those whom men delighted to honour.

The funereal garlands which came, from out of this custom, to give additional solemnity to the beautiful and impressive rites of burial, were sometimes awarded to any who had died in youth, but most generally only to such women who had died virgins. This was another relic of time-honoured usage, which held virginity in peculiar veneration—"out of deference, it would seem, to the Virgin Mother"—and, indeed, in the early Church, almost in sanctity. A special office for virgins is in the Roman Breviary. Widows were permitted to make public vows of celibacy after the death of a first husband, and to these, provided the vows were kept, the honour of these garlands was allowed. They were, like the insignia of a peer or the weapons of a soldier, borne in front of the coffin in the funeral procession, or laid upon it. If a maiden were being buried, two of "the virgins that were her fellows" carried this, the emblem of her purity, before her. Mr. Llewellynn Jewitt quotes an instance in which an old man had carried the garland at the funeral of a young woman of the name of Blackwell, whose garland was suspended from the roof of the church at Ashford-on-the-Water. But it was general for the memento of the departed maiden to be borne by maiden associates. Chambers quotes from *The Virgin's Pattern*—a very scarce little book, which describes the funeral of a lady named Perwich—an account of the manner in which "the trappings of woe" were borne :—

The hearse, covered with velvet, was carried by six servant maidens of the family, all in white. The sheet was held up by six of those gentlewomen in the school that had most acquaintance with her, in mourning habits and white scarfs and gloves. A rich, costly garland of gum-work was borne immediately before the hearse by two proper young ladies that entirely loved her.

The considerable concourse that followed all wore white

gloves. What was done with the "rich, costly garland of gum-work" is not stated, but, most probably, it was, in accordance with the prevailing custom, placed in some part of the parish church, and perhaps, as was usual, over the seat usually occupied by her whose memory it honoured and perpetuated. The token of innocence—as much a trophy of maidenhood as gauntlet and spurs over the tomb of a knight were of valour—was so hung, says Bourne, as "a Token of Esteem and Love, and an Emblem of their Reward in the Heavenly Church;"—

> To her sweet Mem'ry flowery Garlands strung,
> On her now empty seat aloft were hung.
> GAY: *The Garland.*

From the centre of these garlands gloves were suspended, always white, again most probably indicating innocence. First, the wreaths, in direct imitation of the depositary garlands of the ancients, were of myrtle leaves, "most artificially wrought with gold and silver wire," and placed on a hoop of coarse iron wire, covered with a costly fabric of cloth of silver. A particular account of their construction, at a later date, is given in the *Antiquarian Repertory* (*vol. iv.*):—

"The lower rim, or circlet, was a broad hoop of wood, whereunto was fixed, at the sides, part of two other hoops, crossing each other at the top at right angles, which formed the upper part, being about one-third longer than the width. These hoops were wholly covered with artificial flowers, of paper, dyed horn, and silk, and more or less beautiful, according to the skill or ingenuity of the performer. In the vacancy of the inside, from the top, hung white paper, cut in form of gloves, whereon were written the deceased's name, age, &c., together with long slips of

R

various-coloured paper or ribbons. These were many times intermixt with gilded or painted empty shells of blown eggs, as further ornaments, or, it may be, as emblems of bubbles, or bitterness of this life ; whilst other garlands had only a solitary hour-glass hanging therein, as a more significant symbol of mortality."

Sometimes real flowers—lilies and roses—were used, and gloves of kid. But for these, gloves and blossoms both of paper were very generally substituted, and a kerchief of the same material was often added. On kerchief and gloves, not only were the name and age of the deceased inscribed, but in most instances some texts of Scripture or verses, meant to be applicable to the circumstances, and aiming at a moral. These were sometimes chosen for the purpose by the dying maiden herself. By the time that attention was attracted to these pathetic relics the custom had ceased to be observed, and the obliterating hand of time had so far prevailed over the poor ink with which the inscriptions had been written, that they were almost illegible. One, with difficulty deciphered by Mr. Jewitt, ran thus—

> Be always ready—no time delay,
> I in my youth was called away ;
> Great grief to those that's left behind,
> But I hope I'm great joy to find.
>
> ANN SWINDEL,
> Aged 22 years,
> Dec. 9th, 1798.

Another appeared to commemorate Anne Howard, who died at the age of twenty-one, but the six lines of poetry accompanying the record were past reading.

The garlands hung, according to William Howitt, who

wrote of the custom from personal recollection, until they fell through time, or until all who had an interest in the deceased had also gone over to the majority. He mentions that some in Heanor, his native village, disappeared in a general church-cleaning on the advent of a new incumbent who had more love of cleanliness than reverence in his composition. Some found their way, by some means or other, into public or private museums. Others were, in some parts of the country, buried with the maiden, or were taken down on the anniversary of her funeral. Still more, we may suppose, were sacrificed to the practical nature of our ancestors, in whom the bump of veneration could hardly have been very largely developed. In 1662, when an enquiry was made into abuses in the diocese of Ely, one of the interrogatives put was, " Are any garlands, and other ordinary funeral ensigns, suffered to hang where they hinder the prospect, or until they grow foul and dusty, withered and rotten ? "

At Llandovery, it was the practice to take down the garlands after hanging a year, and in the same place each anniversary of death was marked by the decoration of the grave with flowers. On the grave a pair of gloves were placed, which were taken away by the nearest relative who visited it that day.

At one time these garlands were in all probability common to all the churches of the country, for the custom was widely observed. Brand remembered many instances occurring in south-country churches, and others at Walsingham and Stanhope, in the county of Durham, "and in the centre of each a woman's glove cut out in paper" (*Popular Antiquities*). An old inhabitant of Ilkeston, who contributed to a local paper—the *Ilkeston Pioneer*—in 1853, a most interesting series of reminiscences, spoke of counting " more than thirty of these rustic mementoes hanging over the piers." In Derbyshire, where the custom lingered longest, " within the lifetime of those now living," wrote the Rev. J. C. Cox, in his *Churches of Derbyshire*, 1879, such garlands were to be seen within the walls of several churches in the county—Alvaston, Ashover, Bolsover, Eyam, Fairfield, Glossop, Heanor, Hope, Matlock, Tissington, and West Hallam. Anna Seward, a poetess, once known as the Swan of Lichfield, but now well-nigh forgotten, paid a tribute to the garlands in the village church of Eyam—

> Now the low beams, with paper garlands hung
> In memory of some village youth or maid,
> Draw the soft tear from thrilled remembrance sprung ;
> How oft my childhood marked that tribute paid !
>
> The gloves suspended by the garland's side,
> White as its snowy flowers with ribband tied ;
> Dear village ! long may these wreaths funereal spread—
> Simple memorials of the early dead.

A paper in the first number of the *Reliquary* describes, with full details, garlands then extant at Matlock, where six hung from the roof of the vestry, and those at Ashford-in-the-Water, where, in the north aisle, as well as in the chancel of the church at South Wingfield, specimens still remain, and, it is hoped, may long continue. The garlands in general character were all alike. The framework was formed by a broad hoop of wood, to which the segments of two other hoops were attached, crossing each other at right angles at the top, thus forming a kind of open arched crown. The hoops and bands were all of wood, ordinarily of peeled willow, and were wrapped round with white paper, to which were fastened numbers of rosettes, flowers, or pendant ribbons of paper, the latter sometimes "pinked" on the edges.

Funerals are gradually losing the festive element, in which the dead were so often honoured at the expense of the living, and so often dishonoured by unseemly revelry and feasting. The charges incurred on these occasions were excessive, it was a point of pride that the position of the deceased should be upheld by lavish and unstinted hospitality, and an exaggerated estimate of that position, or a degree of such pride, led to a competition which we can now hold in light esteem, even though the feeling which prompts undue display and useless pomp is not extinct even in our day. So, too, the old fashion of making presents of gloves on these occasions yet remains, but so shorn of its former proportions as hardly to be recognized. It was necessary, at one time, to pass an Act of Parliament in Scotland to limit the concourse at a funeral, and such a measure was not altogether unnecessary on this side of the border ; but the gathering at a modern funeral is generally restricted to immediate relatives, so that the gift of gloves to those present does not amount to an inordinate sum. But to show how considerable and heavy a charge this once was, it is only necessary to quote the bill, "In Account of the Funeral of Mary Rudyerd, died in 1717." The total sum of this account is £75 2s. 8d., and the charges include items for the escutcheon, for dinners, wine, and twelve gallons of ale, and for a velvet pall, but out of the aggregate sum 1s. was " Pd. for 2 paer Children's Gloves," and £45 5s. to " Mr. John Sleigh, his Bill for Gloves, Scarves, and Hatbands." In 1639, " Francis Pynner, of Bury, Gent.," left by his will "to every one that shalbe my household servant at the time of my death, twenty shillings a piece, and every one of them a paire of gloves," and another item in this testament desires "Mr. Edmund Callamy that he wilbe pleased to preach at my funerall, and I doe giue him for his paines to be taken therein thirty shillinges and a paire of black gloves of the best sort." In 1664, one David Salter left a sum of two shillings "to be laid out for the buying of a pair of kid gloves, to be given yearly on the first Sunday in Lent to the parson of the parish for the time being," a farm at Thredding Green being charged with the sum.

In giving away gloves, the Universities had no rivals but each other. Any notable personage or royal visitor was welcomed with a present of gloves. The Palatine of Siradia, in 1583, was given some "verie rich and gorgeous gloves." Sometimes the chancellor and the heads of houses, hearing that some persons of consequence were in the neighbourhood, would go out and intercept them to give them gloves. Professor Thorold Rogers, among the muniments of Oxford Colleges, has met with many such instances of gifts of costly gloves, which were presumably not seldom " lined " after the manner objected to by Sir Thomas More. " In 1451, Oriel College gives a pair of gloves to the Bishop of Lincoln, and another to the Chancellor, which cost 2s. 10d. and 2s. 4d. In 1489, it gives a pair at 2s. 10d. to the Bishop of Lichfield ; and in 1504, to the Bishop of Lincoln, at 2s. In 1509, it gives two pairs, at 4s. 4d. and 4s. 6d., to the same prelate. In 1517, Magdalen College gives Wolsey a pair at a cost of 6s. 4d. In 1519, the same College gives two pairs at 2s. 2d., the donee not being named. In 1555, the City of Oxford gives a pair, gold twisted, at a cost of 5s., and another at 2s. 6d. In 1556, Lady Williams, of Thame, has a pair at

4s. In 1559, the City gives a pair to the Earl of Bedford (him of the great head) at 4s. 4d., two pairs to his sons, at 2s. 4d., and 16 pairs to his retinue, at 1s. In 1561, Corpus Christi College gives 4 pairs to Lord and Lady Bedford, at 5s. ; and in 1563, to the clerk of the assize at Winton, at 5s. In 1564, the City gives four pairs to the Judges, at 4s., and to Sir Francis Knollys, Lady Knollys, and the Lord President, at 8s. In 1566, the City gives two pairs to the Lord Admiral and his wife, at 3s. 4d. All Souls gives a pair to the Bishop of Worcester, at 6s. 8d., and to Mr. Lovelace, at 8s. In 1572, Serjéant Lovelace gets another pair from All Souls, at 15s. (here, one can hardly doubt, with a fee inside); Corpus Christi College, three pairs to the Judges and Counsel, at 3s. 3d. ; and two pairs to Sir Walter Mildmay, at 6s. 6d. In 1573, two pairs are presented to the Bishop of Winchester by Corpus Christi College, at 4s. Next year, the same college gives the same prelate a pair, at 5s. In 1575, Magdalen College gives a pair of gloves to Sir Walter Mildmay, at 3s. 4d. In 1576, All Souls gives two pairs to unnamed persons, at 5s.; and Corpus Christi, a pair to the Lord Treasurer at 10s. In 1579, the City of Oxford bestows two pairs of gloves on the Lord Chancellor, each pair costing 10s."

Other instances occur of gifts of gloves to labourers. At Grantchester, in 1436, there is a charge raised of 10s. 6d. for "7 dozen gloves for labourers in autumn ;" and at Fountains, in 1452, for "three dozen gloves," 3s. 4d., with 13 pairs "for Masons," at an average of 2d. per pair. Five years later, at Fountains, is another charge for three pairs of gloves, at 2d. per pair, and a pair at 9d. Magdalen College, in 1556, gives a pair of gloves to some "man" at 10d., and with commendable gallantry a pair "to his wife," of the value of 2s. 2d. In 1494, 4¾d. a pair is charged for "three pair of hedging gloves," and 1½d. a pair "for twelve pairs of harvest gloves." Gloves were given, too, to college tenants—a custom which fell into disuse soon after the reign of Charles I. They were presented in gratitude. When Sir Thomas Pope, in 1556, founded Trinity College, Oxford, the University "complimented him," as well they might, with a letter of thanks, which was accompanied by a gift of rich gloves. (WARTON : *Life of Sir Thomas Pope.*) The gloves were sent both to himself and his lady, and cost 6s. 8d. per pair. On Sir Thomas subsequently visiting Oxford, " the Bursars offered him a present of embroidered gloves." The gratitude of the college was kept up to presentation pitch for a considerable period, even after the death of the founder, for, on the marriage of Sir Thomas's widow with Sir Hugh Powlett, she was presented with a pair of very rich gloves, the charge for which runs, " Pro Pari Chirothecarum dat Dom. Powlett et Domine Fundatrici xvis." The *Progresses* of Nichols shows several instances of gifts of gloves to royal visitors. When James I. and his queen came to Oxford, in 1605, the Chancellor and Bedells of the University presented to his Majesty a Greek Testament in folio, ruled, &c., and two pair of Oxford gloves, with a deep fringe of Gold, the turnovers being wrought with pearle. They cost, as I was informed, £6 a pair. They also gave unto the Queen two pairs of gloves much like the former, and a pair unto the Prince."

Again, when, in 1616, the King was at Woodstock, " the Vice-Chancellor of the University, with certain Heads of Houses, Proctors, and others, went to do their obedience to him, the King receiving them graciously, the Orator made a speech : which being done, the King gave them his hand to kiss, with a promise that he would be favourable to the University, and that learning and learned men should be encouraged. Afterwards they presented to him and certain of the Nobles very rich gloves." The courtiers usually shared in these benefactions. Nichols again writing of a visit of James to Cambridge, in 1622, says : " The University bestowed upon our Chancellor a pair of gloves that cost 44s., and another upon my lord of Waldon of 10s. price. We presented no more in regard, there were so many Lords and great ones of quality. But the next day the two Bishops of London (Mountaine) and Durham (Neile), staying in town all night, the Vice Chancellor and some of the Heads went unto them, and presented them with gloves, about 12s. or a mark a-piece."

These gifts appear to have been carefully proportioned in value to the rank of the recipient, and their worth was made no secret, but rather blazoned abroad. They are not, of course, to be considered disinterested, but mainly given with a purpose to open some negotiations, or introduce some deputation praying a petition. Gloves have always been made to hold this office. In soliciting a favour from some great man, it was usual to break ground by presenting gloves and other valuables to the bigwig's wife. Sometimes the gloves were intended to disarm hostility or invoke intercession. Lysons says that it was usual, when a favour was expected from a minister, to make presents to his lady. In the diary of Alleyne, founder of Dulwich College, occur entries bearing out this conclusion:—

	£	s.	d.
Jan. 1, 1618. Given my lady Clarke a pair of silk stockings......	1	10	0
Given Mr. Austen a pair of silk stockings	1	10	0
Given Mrs. Austen a pair of gloves	1	10	0

The Earl of Hertford writes in 1563, to Lord Robert Duddeley, expressing the grief which he endures by lying under the Queen's displeasure, and soliciting reconciliation, requesting on this behalf that his lordship will present her Majesty "with a poor token of gloves" (*Cal. State Papers: Domestic*). On one occasion, in 1578, the magnates of Cambridge University waited on the Queen at Audley End, on an embassy, which they opened by presenting her with a Greek Testament, and, with some laudatory verses, " a paire of gloves, perfumed and garnished with embroidery, and gouldsmith's wourke, price lxs." "In taking the book and the gloves," continues Nichols, "it fortuned that the paper in which the gloves were folded to open ; and hir Majestie behoulding the beautie of the said gloves, as in great admiration, and in token of hir thankful acceptation of the same, held up one of hir hands, and, then smelling unto them, putt them half waie upon hir hands ; and when the oracon was ended, she rendryed and gave most heartie thanks, promiseing to be mindful of the universitie ; and so, alledging that she was weary, hotte, and fainte aftir hir joyrnie, departed out of the chambre, sending fourthe the aunswere by the lord treasurer, ' That if the Universitie would keepe and perform the promise and condicion made in the oracion, she of hir parte would accomplish their requests and peticion.' After hir Majestie had taken hir chambre, the vice-chauncellour, in the name of the whole Universitie, gave unto the Lord Burleigh, high treasurer of England and lord chancellour, a present of perfumed gloves, price xxs., together with his arms blazd out in colors, with verses annext to them, and a like present with verses to the Earle of Leicester, high steward."

Monarchs sometimes returned these gifts of gloves in

kind. Henry VIII. gave a pair to Sir Anthony Denny, his Privy Councillor and personal friend, whom he appointed executor under his will. These are of leather, the cuff embroidered with silk, gold thread and seed pearls on a satin ground, and fringed with gold and silver lace. To Margaret Edgcumbe, wife of Sir Edward Denny, Elizabeth gave a pair of gloves or mittens of crimson velvet, embroidered with gold and silver thread and silk, with a richly-embroidered cuff of white satin ; and James I. gave to her husband, Sir Edward Denny, afterwards Earl of Norwich, when, in his capacity of Sheriff of Hertfordshire, he received the king during his journey from Scotland, a pair of rich leather gloves, embroidered with gold and silver thread, the cuff decorated with gold and silver lace on crimson silk ground, and fringed. These specimens still exist, and have lately been presented to the South Kensington Museum, by the present Sir Edward Denny, whose ancestor, Sir Thomas Denny, bought them in 1759, at a sale of the Earl of Arran's goods, at a cost respectively of £38 17s., £25 4s., and £22 4s.

CHAPTER IV.

Gloves as Favours.

IN these prosaic days, when our chief boast and aim is Progress, it is not easy to realize the fervour and devotion with which the passion of love was once followed. The sentiment was essentially an unhealthy one—generally unlawful. Where no moral barrier interposed the passion was affected, and, according to our lights, foolish. Indeed, what we know of these rapt devotees reads rather like the most extravagant of burlesques, and what was then a serious and all-absorbing business, would now only be regarded as most exquisite fooling. Yet what here follows is only absolute fact.

The time was that of the early troubadours. For the pleasure and diversion of his kind, when man was kept close confined to his home for the greater portion of the year, and thrown almost wholly on his own resources for amusement, the troubadours sang metrical romances of famous heroes, of their achievements in love and feats in war. They brought to the fireside, where other literature there was none, vivid spectacles of combats, bright pictures of display and pomp, soft episodes of passion, and horrible curdling portraiture of destroying dragons and devouring beasts, who fed on virgins and vanquished knights wholesale. Their hearers in safety enjoyed the luxury of fear which did not frighten, and without trouble enjoyed pleasures as sweet as the world could offer. The troubadours flourished. Poesy became a profession, and to have the gift of narrating tales secured an appointment to the court of a noble, or ensured a wandering minstrel admittance to any house he might seek. There was, of course, rivalry between these masters of verse, and not seldom an appointed tournament of song, when the poets strove with each other for distinction and fame. In all their songs, and particularly before the Crusades altered the course of their romances, the leading motive was Love. All incidents and accidents turned on love, until in time, love, without woman being honoured in it, was made the spring of action and the end of all being. From one degree of excess to another was this philosophy carried, until the tender passion was overloaded with subtle distinctions and refinements, dissected and analyzed until it resembled nothing so much as a dried leaf, which, lacking all beauty and colour, yet could not be robbed of graceful lines and types of eternal truths. From the direct antagonism of troubadours of repute, it is conjectured that Courts of Love came to be established. It is supposed that some of the most prominent of these singers met in a trial of skill before ladies appointed to decide on their respective merits. These Courts of Love afterwards assumed judicial functions in matters of sentiment. Causes were entered before them, and gravely pleaded by suitors, to be decided according to recognized codes of the laws of love, in which were technicalities and definitions and precedents enough to gladden a modern counsel. Love usurped the place of religion, for the Church had become corrupt, and eccle-

siastics were mainly men of the world, holding office under
a lifeless and formal faith.   They, too, were converts to the
new creed, and became passionate pilgrims.   Chaucer's
Monk wears a true-lover's knot, and the Prioress a brooch
bearing an amorous legend.   Thus passion absorbed the
enthusiasm which is generally offered in satisfaction of
spiritual needs, and men and women strove to outdo each
other in excess of devotion to the accepted tenets of
romantic attachments.   A society, styled The Fraternity
of the Penitents of Love, resolutely denied the claims of
nature, and made them subservient to Love.   In summer
they clothed themselves, in proof of their passion and to
show their exaltation above mere feeling, in heavy swelter-
ing furs and thick cloths ; and in winter they wore the
thinnest and most gauzy fabrics, holding it a crime to
shield themselves with gloves or muffs or cloaks, sleeping
under nothing more than a coverlet of thin canvas, and
denying themselves the luxury of warmth in any degree,
so that many of them perished under their voluntary
privations.   One troubadour, becoming passionately en-
amoured of a lady, named Louve de Penantier, styled himself in compliment to her, *Le Loup*, or Wolf, and, to prove
the depth of his devotion, submitted in his simulated character to be hunted by shepherds and their dogs.   Being
overtaken, he was, like Actæon, cruelly mangled by the
hounds, and carried back to his mistress in desperate
plight, but recovered, to be rewarded for his pains by the
lady, who was, as well as himself, glorified in them (*Retrospective Review*).

The pleas raised before the fair judges of the Courts of
Love are amusing enough.   These ladies sat in their robes
of green, with collars of gold, to hear a suitor require that

his elected lady-love should show cause why he should not
be rewarded for his constancy, for the pains he had endured
in pursuing her, making boast of the privations he had suffered, and the risks he had run of being detected by
Danger, as the husband of any lady is generally and very
significantly styled in the pleadings. The defendant would
appear to controvert or deny the facts, or would prove, as
a set-off, that she had, in the risks run on her part, or
anxiety borne during the progress of the affair, more than
counterbalanced any claim the plaintiff might have had on
her pity. Some actions specifically claimed a kiss, some
were for restitution of an alienated affection, or for a com-
pulsion to show politeness. One was brought by a plaintiff
male against a defendant lady for having pricked him
with a pin whilst she was giving him a kiss. The defendant
denied giving the kiss, but said the plaintiff took it, and the
wound, if any, happened by mischance or misadventure.
Evidence was tendered, and certificates of surgeons handed
in to prove the character and extent of the wound; and
the Court finally decided that the defendant should kiss
the wound at reasonable times until it was healed, and,
more practically, even if less effectually, ordered her to
find linen for plaisters.

This amorous pedantry died away. It is believed that
these courts ceased to exist before the close of the thirteenth century. There was, indeed, a later poem entitled
the *Court of Love,* attributed to Chaucer, but it was nothing
more than an embodiment of the institutes of the Love
Court of Provence—the home of troubadours—and contained only the twenty statutes which that court prescribed
to be observed under the severest penalties (WARTON); but
the statutes were then obsolete, and only remained as an

empty ideal to fire the imaginations of poets that came after. In all the metaphysical affection which was then superseded by a more manly and chivalrous devotion to women, we do not know that gloves had any unusual place. Where any trinket or token from a mistress could throw a lover into an ecstacy, it was hardly likely that there should be much difference in degree as to any personal favour. There must have been some limits even to the rapt fervour of professional lovers. But so soon as the wholesome change had set in, and almost all the fantastic ornament with which feeling had been overlaid was stripped away, gloves became an emblem of the newer and purer worship of women. They were made the pledge of affection. The dearest and most cherished of all his array was the little glove which the knight bore proudly in front of his helmet. To scoff at it was a deadly affront, to challenge it the sure commencement of mortal combat. The glove was the heart of honour. It was a constant spur to exertion, an incentive to doughty deeds. It was the loadstone of love : the wearer carried with him a charm against evil, the memory of her who granted it ; he held always the assurance of hope that he would sometime win as his own the hand that fastened it in place, in firm confidence that the wearer would be true to his plighted faith if he lived, and, if he died, would die nobly in defending the honour and trust reposed, through it, in him:

It is this open reverence for woman, this practical testimony to her advancement as she was gradually being lifted from servitude into society, that gave particular odium to the reply of Prince Henry to Earl Percy, when, being told of the justs and tournaments held at Oxford, he boasted that—

> He would unto the stews
> And from the common'st creature pluck a glove
> And wear it as a favour ; and with that
> Unhorse the lustiest challenger.

Said Charles V. once, in like bravado to the French Ambassador, "I have a glove in which I could put your whole city of Paris." Young Henry's answer exemplifies the common confidence in the virtue of the cherished love-gift, at once an amulet and a charm, and by its very audacity shows in how great esteem the favour was then held. .

It is in the sixteenth century that we first meet with the mention of gloves thus worn as favours. Marlowe, in his *Tragedy of Edward II.*, 1598, writes :—

> Nodding and shaking of thy spangled crest
> Where women's favours hang like labels down ;

and in Drayton's *Battle of Agincourt* is a well-known passage :—

> The noble youth, the common rank above,
>   On their courvetting coursers mounted fair.
> One wore his mistress' garter, one her glove,
>   And he a lock of his deir lady's hair :
> And he her colours whom he most did love —
>   There was not one but did some favour wear.

Hall, the chronicler, in a description of a tournament held in the reign of Henry VIII., says of the gallants: "One wore on his head-piece his lady's sleeve, another the glove of his dearlyng," and Sir Thomas Wyat inscribes a short poem—

> *To his love from whom he had her gloues.*
> What nedes these threat'ning wordes, and wasted winde ?
> Al this cannot make me restore my pray,
> To robbe your good, ywis is not my mynde ;
> Nor causelesse your fair hand did I display.

S

Let loue be judge, or els whom next we finde,
That may both heare what you and I can say,
She reft my hart, and I a gloue from her ;
Let us se then, if one be worth the other.

It will be noticed that gloves were not the only favours. All manner of trinkets were mounted aloft as tokens. Mr. Walford, in a remarkable annual, reprints a rare tract of 1594, in the form of a dialogue between the cap and the head, in the course of which the cap complains, "Thou encombrest me with Brouches, Valentines, Ringes, Kayes, Purses, Gloues—yea, fingers of Gloues ; thou wrappest me in Chaines, thou settest me with Buttons and Aglets, thou lardest me with Rybans and Bandes " (*Ephemerides*). Even this list did not exhaust the fantastic ornaments employed in decorating the hat ; for Sir Thomas Overbury describes a courtier and fop "with a pick tooth in his hat."

Sleeves particularly were admitted to rank as favours, while they were yet marks of distinction and honourable. In Shakespeare's version of the most popular tale of the Middle Ages, *Troilus and Cressida*, when the lovers, before parting, change keepsakes, Cressida gives a glove to Troilus and takes from him a sleeve. How she so soon gives this sleeve and her inconstant love to Diomedes, even while she thinks of Troilus, still faithful, giving "memorial dainty kisses" to the glove he holds in pledge, is it not known to every reader of Shakespeare, no less than is the enduring shame which makes the name of Cressida—as she prayed it might be, if she failed her lover—"the crown of falsehood."

Favours were, at the very time when they make a figure in literature, coming into ordinary wear, and from being exclusive were beginning to have an excess of popularity. Spenser, in his pastoral, *Shepheard's Calendar*, makes Cuddie say to Thenot that, if his years were green, instead of sere and yellow, if he had not outlived passion, he would learn—

> To caroll of Love,
> And hery (praise) with hymnes thy lasses glove.

The universal adoption of favours led to their ruin. Already there was about the custom a gathering suspicion of effeminacy. In Lyly's *Alexander and Campaspe*, Parmenio complains to Clytus, " Thy men are turned to women, thy soldiers to lovers, gloves worn in velvet caps, instead of plumes in graven helmets." A little later and they were worn by domestics and retainers, and their downfall was complete. Edgar, in *Lear*, says that he has been "a serving-man proud in heart and mind, that curled my hair, wore gloves in my cap;" and in Beaumont and Fletcher's *Scornfull Ladie*, 1616, Welford being offered a glove as a favour by Abigail, replies, "Harke you, mistres, what hidden virtue is there in this glove, that you would have me weare it? Is 't good against sore eyes, or will it charm the toothache? Or, are these red tops, being steept in white wine, will't cure the itch? Or has it so concealed a providence to keep my hand from bonds? If it have none of these and prove no more but a bare glove of half-a-crowne a paire, 'twill be but half a courtesy. I wear two always."

A fall indeed was this from the time, not long before, when they were given by a queen as a mark of her favour! "Here Glisson, wear them for my sake," said Elizabeth to Dr. Glisson, pulling from her hands a pair of gloves, which, as the recipient tells us, were of rich Spanish leather, embossed on the backs and tops with gold embroidery, and fringed round with gold plate. (*Life of Corinna*). Elizabeth, on another occasion, gave a glove to Clifford, Earl of Cumberland, when he brought it back after she had dropped it. He was told to keep it as a mark of his sovereign's esteem, and he had it right royally adorned, wearing it always in front of his hat in tournaments. (*Bray's Tour.*) The gage was probably dropped coquettishly. Strong as was Elizabeth's hand, she hid it beneath the velvet glove, and ruled, however firmly, still with sentiment and passion. She ought never to have been a woman, but being a woman she governed by a woman's artifices and strategy. Her ministers were her lovers, we are told, and marriage with her was a prize to be diplomatically dangled before many suitors. She had, doubtless, her moments of unfeigned affection, the only difficulty being to distinguish between the simulated and the real inclination. There is a story told of her dropping a glove on the stage before Shakespeare, on one of the several occasions on which his company acted before the Court. The poet is said to have been taking the part of a king. He had doubtless plenty of enthusiasm in his art—may, perhaps, have been playing in one of his own tragedies, and lost himself in the character. The presence and applause of the Queen would stimulate him; and he was, we may believe, a man of comely presence; and Elizabeth liked to see handsome men, and had a shuddering horror of deformity. In any case, as the story goes, he acted so much to her Majesty's satisfaction, that she threw her glove to him on the stage—a challenge, maybe, to a flirtation. But, with a ready wit and wise discretion, the poet affected to believe it an accident, picked up the glove, and returned it to the Queen, saying :—

> Although now bent on this high embassy,
> Yet stoop we to pick up our Cousin's glove.

When the wearing of favours had become common, and, indeed, to be considered somewhat foolish, Ben Jonson ridicules the fading custom in broadest farce :—

> Marry, some houre before shee departed, shee bequeath'd to me this glove, which golden legacie, the Emperour himselfe tooke care to send after me, in six coaches, cover'd all with black vellvet, attended by the state of his empire ; all which he freely presented me with, and I reciprocally (out of the same bountie) gave to the lords that brought it ; only reserving the gift of the deceas'd lady, upon which I compos'd this *Ode*, and set it to my most affected instrument, the Lyra.

> Thou more than most sweet Glove,
> Unto my more sweet love,
> Suffer me to store with kisses
> This emptie lodging, that now misses
>   The pure rosie hand that ware thee,
>   Whiter than the kid that bare thee,
>   Thou art soft, but that was softer ;
>   CUPID's selfe hath kist it ofter,
>   Than e'er he did his mother's doves,
>   Supposing her the Queen of loves,
>   That was thy mistresse,
>   Best of Gloves. —*Cynthia's Revels.*

The dandies "in the reign of good Queen Anne" would have revived this fashion of personal favours, but certainly not in the high spirit of chivalry. "Instead of snuff-boxes and canes," says the *Spectator*, "which were usual helps to discourses with other young fellows, these have each some piece of riband, or broken fan, or an old girdle which they play with while they talk of the fair person remembered by each respective token."

Here ends the chronicle of gloves, and here we may well write ICHABOD, for the glory of gloves is departed. The times are changed and we with them. Still, there is nothing lost in recalling the days when gloves were something more than a mere covering for the hands, when they were an outward and visible sign of faith in woman and the touchstone of honour in man.

For the effort to set forth the traditions or trace the annals of an article which we still wear, but, like rue, with a difference, there is, in these days of archæological study, happily no need to attempt palliation; nor is excuse required for the endeavour to show how an article, apparently commonplace and of little worth, may be made to stand representative of associations, and in some measure illustrative of incidents which make up our history. In the close scrutiny of the past, which is so rapidly gaining in favour, gloves must necessarily have a place. In the science of archæology which is doing so much to make history more than a dry record of events— which is endowing our forefathers with new life and presenting them not so much as characters in fiction, but as our very relatives, living as ourselves—in this science gloves must claim a part commensurate with their former importance.

Archæology, although so popular, is comparatively young. The Society of Antiquaries was truly founded so long ago as 1572, but it cannot be pretended that the study of antiquity is of more than modern growth. Indeed, it is only within the past few years that the subject has been really studied at all. Jonathan Oldbuck was only an antiquary in a limited sense, and his race is extinct. In place of ponderous treatises, replies, rejoinders, and replications, on relics of dubious value and of indeterminate age, we have now veritable and authentic records, lovingly collected and carefully studied ; we have ardent explorers unearthing the memorials of early races on every hand ; we have diligent collectors of folk-lore, and students of dialects—all doing their work with unwearied zeal, and very generally without hope of any other reward than that of having done it well. Learned societies, devoted to the elucidation of history, abound and flourish exceedingly. Antiquarian publications are multiplying to meet a fast-growing appetite ; newspapers in every direction are establishing or carrying on well-established columns of Notes and Queries illustrative of local history. National records are being laid open to all who care to read them, and private collections giving freely to the public all their treasures. Details of the domestic life of our forefathers are becoming essential features of school primers, and we have even had a novel published with the avowed idea, not so much of presenting an attractive story, as of conveying a picture of sixteenth-century England.

Every part of this very general study reacts on every other, and it would be hopeless to expect that the last word had been written even in a History of Gloves. But, at all events, something has been done to establish the dignity of gloves, to show their long descent and their value in costume, and to give them the position in the history of antiquity, to which their intimate relationship with the affairs of men fairly entitles them.

Printed by W. H. and L. COLLINGRIDGE, 148 and 149, Aldersgate Street, E.C.